软件工程技术专业系列教材

资源分享应用后台管理系统开发实战

徐逸卿 刘 波 主编

许莫淇 孙宇航
金晨星 王 辰　参编

东南大学出版社
SOUTHEAST UNIVERSITY PRESS
·南京·

图书在版编目(CIP)数据

资源分享应用后台管理系统开发实战 / 徐逸卿,刘波主编. -- 南京：东南大学出版社, 2024. 12.
ISBN 978-7-5766-1711-5

Ⅰ. TP312.8

中国国家版本馆 CIP 数据核字第 20248QN033 号

责任编辑：夏莉莉　责任校对：韩小亮　封面设计：毕　真　责任印制：周荣虎

资源分享应用后台管理系统开发实战

Ziyuan Fenxiang Yingyong Houtai Guanli Xitong Kaifa Shizhan

主　编	徐逸卿　刘　波
出版发行	东南大学出版社
出 版 人	白云飞
社　　址	南京市四牌楼 2 号　邮编：210096
网　　址	http://www.seupress.com
经　　销	全国各地新华书店
印　　刷	广东虎彩云印刷有限公司
开　　本	787 mm×1092 mm　1/16
印　　张	17.25
字　　数	455 千字
版　　次	2024 年 12 月第 1 版
印　　次	2024 年 12 月第 1 次印刷
书　　号	ISBN 978-7-5766-1711-5
定　　价	60.00 元

本社图书若有印装质量问题,请直接与营销部联系。电话(传真):025-83791830

目 录

故事的开始 ……………………………………………………………………………… 1

准备任务　了解项目 ……………………………………………………………… 3
 1　初识项目 ………………………………………………………………………… 3
  1.1　任务描述 …………………………………………………………………… 3
  1.2　任务分析 …………………………………………………………………… 3
 2　项目架构设计 …………………………………………………………………… 9
  2.1　任务描述 …………………………………………………………………… 9
  2.2　任务分析 …………………………………………………………………… 9
 3　项目技术选型 …………………………………………………………………… 14
  3.1　任务描述 …………………………………………………………………… 14
  3.2　任务分析 …………………………………………………………………… 14
 4　项目开发规划 …………………………………………………………………… 18
  4.1　任务描述 …………………………………………………………………… 18
  4.2　任务分析 …………………………………………………………………… 18
 5　任务总结 ………………………………………………………………………… 20

任务一　设计数据库 ……………………………………………………………… 21
 1　数据库设计 ……………………………………………………………………… 21
  1.1　任务描述 …………………………………………………………………… 21
  1.2　任务分析 …………………………………………………………………… 22
 2　数据表结构设计 ………………………………………………………………… 26
  2.1　任务描述 …………………………………………………………………… 26
  2.2　任务分析 …………………………………………………………………… 26
 3　数据库建表脚本 ………………………………………………………………… 37
  3.1　任务描述 …………………………………………………………………… 37
  3.2　任务分析 …………………………………………………………………… 37
 4　任务总结 ………………………………………………………………………… 45

任务二　设计接口文档 ·· 47
1　后台管理系统接口文档 ·· 47
1.1　任务描述 ·· 47
1.2　任务分析 ·· 48
2　任务总结 ··· 52

任务三　搭建后台管理系统 ·· 54
1　后台管理系统后端 ··· 54
1.1　任务描述 ·· 54
1.2　任务分析 ·· 55
2　文件上传 ··· 64
2.1　任务描述 ·· 64
2.2　任务分析 ·· 65
3　后台管理系统前端 ··· 73
3.1　任务描述 ·· 73
3.2　任务分析 ·· 73
4　任务总结 ··· 83

任务四　用户管理模块开发 ·· 85
1　用户列表 ··· 85
1.1　任务描述 ·· 85
1.2　任务分析 ·· 86
2　查看和编辑用户信息 ·· 100
2.1　任务描述 ·· 100
2.2　任务分析 ·· 100
3　导出用户 ··· 109
3.1　任务描述 ·· 109
3.2　任务分析 ·· 110
4　冻结用户 ··· 115
4.1　任务描述 ·· 115
4.2　任务分析 ·· 116
5　任务总结 ··· 119

任务五　积分管理模块开发 ·· 120
1　后端基础类创建 ·· 120
1.1　任务描述 ·· 120
1.2　任务分析 ·· 121

2 积分管理模块前端实现 ·· 124
2.1 任务描述 ·· 124
2.2 任务分析 ·· 125
3 任务总结 ·· 133

任务六 通知管理模块开发 ·· 134
1 通知管理模块后端实现 ·· 134
1.1 任务描述 ·· 134
1.2 任务分析 ·· 135
2 通知管理模块前端实现 ·· 142
2.1 任务描述 ·· 142
2.2 任务分析 ·· 142
3 任务总结 ·· 155

任务七 标签管理模块开发 ·· 157
1 标签管理模块后端实现 ·· 157
1.1 任务描述 ·· 157
1.2 任务分析 ·· 158
2 标签管理模块前端实现 ·· 165
2.1 任务描述 ·· 165
2.2 任务分析 ·· 166
3 任务总结 ·· 178

任务八 分类管理模块开发 ·· 179
1 分类管理模块后端实现 ·· 179
1.1 任务描述 ·· 179
1.2 任务分析 ·· 180
2 分类管理模块前端实现 ·· 185
2.1 任务描述 ·· 185
2.2 任务分析 ·· 186
3 任务总结 ·· 197

任务九 资源管理模块开发 ·· 199
1 资源列表 ·· 199
1.1 任务描述 ·· 199
1.2 任务分析 ·· 200
2 资源审核 ·· 214

2.1　任务描述 ·· 214
　　2.2　任务分析 ·· 215
　3　补充积分业务 ·· 221
　　3.1　任务描述 ·· 221
　　3.2　任务分析 ·· 222
　4　任务总结 ·· 224

任务十　首页仪表盘开发 ·· 225
　1　仪表盘效果展示 ·· 225
　　1.1　任务描述 ·· 225
　　1.2　任务分析 ·· 225
　2　首页仪表盘后端实现 ·· 226
　　2.1　任务描述 ·· 226
　　2.2　任务分析 ·· 227
　3　首页仪表盘前端实现 ·· 231
　　3.1　任务描述 ·· 231
　　3.2　任务分析 ·· 231
　4　任务总结 ·· 250

任务十一　后台管理系统打包部署 ·· 251
　1　后端环境搭建 ·· 251
　　1.1　任务描述 ·· 251
　　1.2　任务分析 ·· 252
　2　后端 API 部署 ··· 257
　　2.1　任务描述 ·· 257
　　2.2　任务分析 ·· 258
　3　前端部署 ·· 260
　　3.1　任务描述 ·· 260
　　3.2　任务分析 ·· 261
　4　任务总结 ·· 265

结束任务　客户端系统 ·· 266
　1　客户端整体功能 ·· 267
　2　客户端技术选型 ·· 267
　3　项目开发规划 ·· 268
　4　任务总结 ·· 269

故 事 的 开 始

在数字化时代的浪潮中,资源分享应用的崛起成为互联网领域的一大亮点。资源分享应用不仅为用户提供了便捷的资源获取途径,还极大地促进了知识的传播与共享。在这样的背景下,小南和小工,两位刚从大学毕业的热血青年,怀揣着对技术的热爱和对未来的憧憬,踏入了位于科技园区内的一家充满活力的软件开发公司。

初入公司

入职公司的第一天,小南和小工就被公司充满活力与创新的工作氛围所吸引。他们迫不及待地想要开始自己的职业生涯,并期待能够在软件开发领域大展拳脚。在简短的入职培训后,他们得知公司正在计划开发一款全新的资源分享应用,并且急需后台管理系统的支持。作为新人,他俩有幸被选中参与这个项目,成为这个重要项目的一员。

小南的忐忑

虽然小南对编程有着浓厚的兴趣,但面对实际的项目开发,他心中还是充满了忐忑。他担心自己的技能水平不够,无法胜任这个重任;害怕在项目中犯错,给团队带来麻烦。但是,他也知道这是一个难得的机会,必须勇敢地面对挑战,克服自己的恐惧。

小工的自信与担忧

与小南不同,小工在大学期间积累了一定的项目经验,对于后台管理系统的基本概念和技术有一定的了解。他对于能够参与这个项目感到兴奋,但同时也担心自己的知识储备是否能够应对项目中的各种复杂问题。他明白,这个项目不仅是一个技术挑战,更是一个团队协作和成长的机遇。

项目的启动会议

在项目经理的召集下,小南和小工参加了项目的启动会议。在会议上,他们详细了解了项目的背景、目标、时间计划和团队分工。项目经理向他们介绍了项目的整体架构和设计理念,并强调了后台管理系统在项目中的重要性。项目经理告诉他们,这个项目将采用任务化的方式进行开发,每个阶段都有明确的任务目标和时间节点。

导师的悉心指导

为了帮助他们更好地完成项目任务,公司为他们配备了一位经验丰富的导师:波哥。波哥不仅会为他们提供技术上的指导,还会帮助他们分析业务需求,引导他们如何设计出一个既满足业务需求又易于维护的后台管理系统。波哥耐心地解答他们的疑惑,给予他们鼓励和支持,让他们逐渐克服了初入项目时的迷茫和不安。

章节预告

在接下来的章节中,我们将以小南和小工的成长为主线,带领大家一步步完成这个资源分享应用后台管理系统的开发。每个章节都将围绕一个具体的任务展开,包括了解项目、设计数据库、设计接口文档、搭建后台管理系统、开发各个功能模块等。在每个任务中,小南和小工都会面临一些挑战和困难,但是他们通过不断的学习和实践,最终成功地完成了任务。在这个过程中,他们不仅提升了自己的编程能力,还学会了如何与团队成员有效沟通协作。让我们一起跟随小南和小工的脚步,开始这段充满挑战和成长的旅程吧!

准备任务

了 解 项 目

清晨的阳光透过办公室的窗户,洒在小南和小工的脸上,今天是他们参与资源分享应用后台管理系统开发工作的第一天。对于这个全新的挑战,小南和小工既兴奋又紧张。项目经理召集小南和小工参加了项目启动会,向他们介绍了项目的整体情况和开发规划,并安排了他们的第一个任务:了解项目。

◇ 任 务 点

- 了解项目的整体功能;
- 知道最新主流的项目技术选型,理解本项目的技术选型;
- 了解项目开发规划。

◇ 任 务 计 划

- 任务内容:了解项目的功能、技术选型和开发规划等;
- 任务耗时:预计完成时间为 30 min~1 h;
- 任务难点:分析项目的技术选型。

1 初识项目

1.1 任务描述

项目启动会之后,小南和小工依然没有方向,就追着波哥问:"波哥,马上就要开始做项目,我们应该做些什么准备工作呢?"

波哥语重心长地说:"知己知彼,才能百战百胜啊。我们先要了解什么是资源分享应用后台管理系统,然后进行后台管理系统的整体规划、设计整个系统的业务功能和业务流程、构思好项目的总体架构、选择合适的开发技术,这样才能在开发时游刃有余。"

1.2 任务分析

1.2.1 认识资源分享应用后台管理系统

在数字化、信息化快速发展的今天,合作与互助的精神愈发显得重要。资源分享,作为合作与互助的一种具体体现,已经渗透到社会的各个角落,包括教育、科研、企业等多个领域。它不仅促进了资源的优化配置和知识的广泛传播,还加强了人与人之间的联系和信任,对推动社会的进步与发展起到了至关重要的作用。因此,开发一个高效、便捷的资源分享应用,并配备强大的后台管理系统,已成为满足时代发展需求、推动社会进步的重要措施。

1. 资源分享的意义与价值

资源分享,指的是个体或组织将其所拥有的资源与他人共享的行为。这些资源可以是物质资源,如图书、设备、场地等;也可以是非物质资源,如知识、经验、技能等。通过资源分享,不仅可以实现资源的最大化利用,还可以促进知识的广泛传播和人际关系的和谐发展。

在教育领域,资源分享尤为重要。教师可以通过分享教学资源,如教案、课件、教学视频等,为学生提供更加丰富、多样的学习资源,使学习更加便捷、高效。而学生之间也可以互相分享学习资料、学习方法和经验,从而促进彼此的学习和交流。此外,资源分享还有助于培养学生的合作精神和分享意识,为未来的职业发展和社会适应打下良好的基础。

在企业领域,资源分享同样具有重要意义。员工之间可以通过分享工作资料、经验和技能等,提高工作效率和团队协作能力。这不仅可以减少重复劳动和浪费,还可以促进知识的积累和传承,为企业的发展提供有力支持。此外,企业还可以通过资源分享与外部合作伙伴建立更紧密的联系,实现资源共享和互利共赢。

在科研领域,资源分享同样发挥着重要作用。研究人员可以通过分享数据、实验成果等资源,促进科研信息的交流和合作。这不仅可以加快科研进程,还可以避免重复研究和资源浪费。同时,资源分享还有助于建立科研诚信和合作文化,为科研事业的长期发展奠定坚实基础。

2. 资源分享应用的开发背景

随着信息时代的到来,资源与信息的丰富多样性使得人们急需一个高效、便捷的平台来查找、获取和分享资源。传统的资源分享方式,如面对面交流、纸质资料传递等,已经无法满足人们的需求。因此,开发一个资源分享应用成为满足时代发展需求并推动社会进步的重要措施。

资源分享应用具有以下优势:

- 高效便捷:用户可以通过手机或电脑等设备随时随地访问应用,查找和获取所需资源。同时,应用还提供了搜索、分类等功能,方便用户快速找到所需资源。
- 资源丰富:应用汇聚了来自各个领域的资源,包括教育、科研、企业等多个领域。用户可以根据自己的需求和兴趣,轻松获取各种类型的资源。
- 互动性强:用户可以在应用内与其他用户进行交流和互动,分享自己的资源和经验。这种互动不仅有助于促进知识的传播和共享,还可以增强用户之间的联系和信任。

3. 后台管理系统的功能与特点

在资源分享应用中,后台管理系统扮演着至关重要的角色。它不仅是整个应用的核心支撑,还是确保应用稳定、高效运行的关键所在。以下将对后台管理系统的功能与特点进行详细解析。

- 资源管理功能

后台管理系统能够高效地管控平台上的各类资源。通过资源上传、审核、发布等功能,确保资源的合法性和合规性,维护良好的内容生态。同时,系统还提供了资源分类、标签等功能,方便用户快速找到所需资源。此外,后台管理系统还能够对资源的使用情况进行监控和分析,为资源的优化和更新提供数据支持。

- 用户管理功能

后台管理系统能够对用户进行有效管理,保障平台的安全性和秩序性。通过用户注册、登录、权限管理等功能,确保用户身份的真实性和合法性。同时,系统还提供了用户行为监控、数据分析等功能,帮助管理员了解用户需求和行为特点,为平台的优化和发展提供数据支持。此外,后台管理系统还能够根据用户的反馈和投诉,及时处理问题并改进服务。

- 数据分析功能

后台管理系统具备强大的数据分析功能。通过收集和分析用户行为数据、资源使用情况等数据,深入了解用户需求和资源使用情况。这些数据不仅可以为平台的优化和发展提供有力支持,还可以为运营策略的制定和合作伙伴的选择提供数据依据。同时,数据分析功能还有助于发现潜在的问题和风险,为平台的安全稳定提供有力保障。

- 系统监控与故障排查功能

后台管理系统能够对系统进行实时监控和故障排查。通过监控系统的运行状态、资源使用情况等指标,及时发现潜在的问题和风险。当系统出现故障时,后台管理系统能够迅速定位问题所在并及时进行修复,确保平台的稳定运行。此外,系统监控功能还能够为管理员提供实时预警和提醒服务,确保管理员能够及时发现异常情况。

- 版本更新与维护功能

随着技术的不断发展和用户需求的不断变化,资源分享应用需要不断进行更新和维护。后台管理系统能够提供版本更新和维护功能,确保应用能够持续满足用户需求并保持良好的运行状态。通过定期发布新版本、修复已知问题、添加新功能等方式,不断提升用户满意度。

1.2.2 使用人群和主要功能分析

在构建一个高效且用户交互友好的后台管理系统前,对目标用户群体的深入理解和角色划分至关重要。这些用户角色不仅代表了不同的使用场景和需求,还决定了系统应具备的功能和权限设置。以下是对资源分享应用后台管理系统的主要用户角色及其所需功能的详细分析。

1. 系统管理员

系统管理员是整个后台管理系统的核心角色,负责平台的全面管理和监督。其主要职责和功能包括：
- 平台管理：拥有对系统所有功能的完全访问权限，可以监控整个平台的运行状态，进行必要的系统配置和调整。
- 用户管理：能够添加、修改、删除和禁用用户账户，设定用户的访问权限和角色。
- 内容审核：负责审核用户上传的资源，确保内容符合法律法规和平台政策，维护平台的内容质量和安全性。
- 问题处理：处理系统级别的问题，如安全性问题、数据泄露等，确保平台的稳定运行。
- 系统更新：管理和发布系统更新，修复潜在的安全漏洞和性能问题，提升用户体验。

2. 内容提供者

内容提供者通常是教师、研究人员、业界专家等，他们为平台提供高质量的专业内容或资源。其主要功能包括：

- 资源上传：能够上传自己的资源，如教学视频、研究报告、专业文章等，并设置资源的访问权限和分享方式。
- 资源管理：对上传的资源进行管理，包括进行修改、更新、删除等操作，确保资源的准确性和时效性。
- 用户反馈处理：接收和处理用户对资源的反馈，根据反馈信息修改或更新资源内容，提升资源质量。

3. 数据分析师

数据分析师负责分析应用数据，为平台的发展提供数据支持。其主要功能包括：

- 数据分析：分析用户行为、资源流行度、下载量等数据，了解用户需求和资源使用情况。
- 功能优化：根据数据分析结果，提出改进平台功能和内容布局的建议，提升用户体验。
- 报告生成：生成详细的数据报告，为管理层提供决策支持，帮助平台更好地发展。

4. 客户端用户（非后台管理角色）

虽然客户端用户（如学生、企业员工、自学者等）不是后台管理系统的直接用户，但他们的需求和使用行为对系统设计和功能设置有着重要影响。系统需要满足这些用户的需求，提供以下功能：

- 资源浏览与搜索：提供直观的资源浏览界面和强大的搜索功能，帮助用户快速找到所需资源。
- 资源下载与分享：允许用户下载所需资源，并支持将资源分享给他人，扩大资源的传播范围。
- 个性化推荐：根据用户的兴趣和行为，推荐相关资源和内容提供者，提升用户体验。
- 用户反馈：提供用户反馈渠道，收集用户对资源和平台的意见和建议，帮助平台不断改进和优化。

通过以上角色划分和功能分析,我们可以更好地设计出满足各种用户需求的功能和权限设置,确保资源分享应用后台管理系统的高效、稳定和易用。在本书的后台管理系统中,主要用户角色仅包括管理员和数据分析师,其余角色属于客户端用户。通过这样的角色划分,可以更好地设计出满足各种用户需求的功能和权限设置。

1.2.3 角色分析与功能说明

经过波哥的介绍,小南和小工终于对资源分享应用后台管理系统的整体背景有了初步了解。

小南开始研究后台管理系统的功能,了解到本系统涵盖了基础功能模块、用户管理模块、积分管理模块、通知管理模块、资源管理模块和数据大屏模块等多个方面(表1)。这些模块共同构成了强大的后台管理功能,实现了用户对系统进行全面的管理和监控。

表1 后台管理系统的主要模块及其功能

模块	功能
基础功能模块	RBAC(Role-Based Access Control,基于角色的访问控制)权限:角色管理、菜单管理、授权管理等
用户管理模块	用户信息查看、禁用/启用
积分管理模块	积分查询
通知管理模块	通知增删改查、置顶
资源管理模块	资源查看、审核;标签增删改查
数据大屏模块	提供标签词云图、资源类型占比饼图、用户总数、今日签到总数、资源总数等数据统计

1.2.3.1 用户角色分析

1. 系统管理员

系统管理员是后台管理系统的核心角色,负责整个平台的日常运营和维护。其主要职责包括用户管理、资源审核、通知发布以及系统的整体安全维护。

- 功能说明
  - 基础功能模块:管理员可以定义和管理不同的用户角色,为每个角色分配特定的权限集合;设置和管理系统菜单,确保不同角色的用户只能访问其权限范围内的功能。
  - 用户管理模块:管理员可以查看所有用户的详细信息,包括修改和导出用户的基础信息;根据平台政策和用户行为,禁用或启用用户账号。
  - 资源管理模块:管理员负责审核所有上传的资源,确保内容的合法性和适宜性;同时,可以增删改查资源的标签和分类,帮助优化资源的分类和检索。
  - 通知管理模块:管理员可以发布、修改、删除和置顶各种通知内容,确保用户及时获取重要信息。

- 数据大屏模块：管理员可以访问并分析各种数据，如标签词云图、资源类型占比饼图、用户总数、今日签到总数、资源总数等，为平台运营提供数据支持。

2. 数据分析师

数据分析师主要负责数据监控和分析，通过深入解读数据来优化用户行为，了解资源使用情况以及平台的整体表现。

- 功能说明
  - 数据大屏模块：数据分析师可以访问并分析与平台运营相关的各种数据，如用户活跃度、资源下载量、用户反馈等。同时，利用数据分析工具支持决策制定，提出改进建议。

3. 普通用户

普通用户是后台管理系统的核心服务对象，虽然不直接参与后台管理，但其使用行为是系统运营的基础。普通用户角色在后台管理系统中主要体现为"被管理对象"。

- 功能说明
  - 基础信息维护：可查看、修改个人基本信息（如用户名、联系方式等）。当修改敏感信息时，需通过管理员审核。
  - 资源交互功能：具有资源上传入口，可提交待审核资源；可查看个人资源审核状态及历史记录；可对已发布资源进行下架申请。
  - 积分系统使用：可实时查询个人积分余额及明细；可查看积分获取/消耗规则说明。
  - 通知接收：可查看系统公告和个性化通知；具备通知已读状态标记功能；支持通知内容反馈。

1.2.3.2 功能模块详解

1. 基础功能模块

基础功能模块是管理系统的核心，它基于 RBAC 模型，实现用户角色的定义、权限分配和菜单管理。通过这一模块，管理员可以灵活配置系统权限，确保系统的安全性和易用性。

2. 用户管理模块

用户管理模块是系统的重要组成部分，它负责维护用户的信息。管理员可以通过该模块查看、修改、导出用户的基础信息，并根据平台政策和用户行为禁用或启用用户账号。

3. 积分管理模块

积分管理模块用于激励和评价用户的参与度和活跃度。管理员可以查看用户的积分增减记录，但不能直接修改积分数值。这一模块可以有效地鼓励用户更多地参与平台活动，提升用户的参与度和归属感。

4. 通知管理模块

通知管理模块是维护与用户沟通的关键组件。管理员可以通过该模块发布、修改、删除和置顶各种通知内容，确保用户及时获取重要信息。同时，重要通知可以设置置顶或首页轮播，提高用户的关注度。

5. 资源管理模块

资源管理模块是确保资源高效管理和发布的核心部分。管理员可以通过该模块审核和查看用户上传的资源,确保内容的合法性和适宜性。同时,管理员还可以增删改查资源的标签和分类,帮助优化资源的分类和检索。

6. 数据大屏模块

数据大屏模块是管理系统中用于实时数据展示和分析的关键部分。它提供多种可视化组件,如图表、地图、表格等,帮助管理员和数据分析师直观地监控、分析和理解各种关键指标和数据趋势。这一模块为平台的运营和优化提供了有力的数据支持。

2 项目架构设计

2.1 任务描述

波哥仔细审视着项目文件,深知新项目的重要性,于是召集了小南和小工来讨论项目的架构设计。

他强调,新项目的成功与否,很大程度上取决于能否设计出一个合理且高效的架构。架构设计是项目成功的基石,需要综合考虑技术、需求和团队能力等多方面因素,制定出一个合适的方案。

在讨论过程中,小工提出了他的初步想法:"我认为我们首先需要明确项目的需求和目标,这是选择合适架构模式的前提。例如,如果项目对性能有较高要求,我们可以考虑采用微服务架构;若更注重快速迭代和扩展性,那么基于事件驱动的架构可能更为合适。"

小南也补充说:"除了技术和团队因素外,我们还需要考虑一些非技术因素,如项目的预算、时间限制以及可能的法律法规要求。这些因素同样会对架构设计产生重要影响。"

波哥对两人的观点表示赞同,并建议他们围绕这些要点展开讨论,以设计出一个既合理又切实可行的项目架构。在波哥的指导下,团队成员们开始围绕项目的架构设计展开讨论,并着手制定详细的实施方案。

2.2 任务分析

资源分享应用后台管理系统作为支持在线资源共享和管理的关键平台,其系统架构的设计和实现显得尤为重要。系统架构不仅决定了系统的功能性表现,还直接关系到系统的非功能性需求,如性能、可扩展性、可靠性、安全性等。因此,对系统架构进行深入分析和合理选择,对于确保系统的稳定、高效运行至关重要。

系统架构是对系统中实体以及实体之间关系所进行的抽象描述。它旨在实现业务功能的同时,满足非功能性需求。系统架构的设计需要综合考虑业务需求、系统规模、技术团队实力等因素,以确保系统的稳定、高效运行。

不同的架构风格在支持业务功能方面可能相似,但在非功能性表现上则各有千秋。例如,单体架构在开发初期具有结构简单、开发成本低等优点,但随着系统规模的扩大和业务复杂度的增加,其维护困难、扩展性差等问题逐渐凸显;分布式架构通过垂直切分将系统拆分为多个子系统,提高了系统的可扩展性和可靠性,但也带来了系统间耦合度高、数据冗余等问题;SOA架构和微服务架构则通过服务拆分和接口定义,实现了服务的重用和集成,提高了系统的可重用性和可维护性。而服务网格则通过在网络层实现服务治理和通信,进一步降低了服务间的耦合度,提高了系统的灵活性和可扩展性。

2.2.1 单体架构

单体架构是Web应用开发初期阶段的一种常见架构模式。它将项目的所有业务功能模块都打包放在一个Web容器中运行,并使用同一个数据库。这种架构具有结构简单、开发成本低、效率高等优点,适用于小型项目的快速开发。随着系统规模的扩大和业务复杂度的增加,单体架构将面临以下挑战:

- 维护困难:随着代码量的增加,系统的复杂性逐渐提高,维护成本也随之增加。
- 扩展性差:单体架构将所有功能都集成在一个应用中,导致系统难以水平扩展。
- 数据库压力大:所有服务都使用同一个数据库,导致数据库压力大,性能瓶颈明显。
- 技术栈单一:单体架构往往采用统一的技术栈,限制了新技术的引入和应用。

单体架构的结构如图1所示。

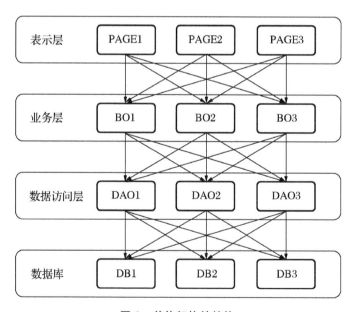

图1 单体架构的结构

2.2.2 分布式架构

随后出现了分布式架构,也称为垂直架构。顾名思义,它将单体架构按业务功能垂直切分,拆分后的系统之间通过网络进行交互,并且每个系统都可以独立部署。通过对服务的垂直切分,每个子系统的业务功能简单,开发周期短,更加高效灵活。但是这样的拆分方式会导致系统和系统之间耦合度过高,数据冗余,会降低系统的可靠性,并且实现的技术复杂。

分布式架构的优点包括:
- 可扩展性强:通过水平扩展子系统,可以满足系统不断增长的业务需求。
- 可靠性高:每个子系统都独立运行,某个子系统的故障不会影响整个系统的运行。
- 技术栈灵活:不同子系统可以采用不同的技术栈,有利于新技术的引入和应用。

分布式架构也带来了以下挑战:
- 系统间耦合度高:由于子系统之间存在数据交互和业务依赖,导致系统间耦合度高。
- 数据冗余:不同子系统之间可能存在数据冗余,增加了数据维护的复杂性和成本。
- 技术复杂度高:分布式架构需要处理跨节点通信、数据一致性等问题,技术实现复杂度较高。

2.2.3 SOA(Service-Oriented Architecture)架构

SOA 是一种基于分布式架构思想的面向服务的架构,它可以动态地改变用户以及应用程序对信息的需求。SOA 根据系统的不同业务进行服务拆分,通过这些服务中预定义的接口进行通信。SOA 将重复的功能抽离成公共组件向每个系统提供服务,提高了开发效率和系统的可重用性。但是由于服务间通信没有一个标准,导致服务间的接口协议不固定,不利于系统的维护。

SOA 架构的优点包括:
- 可重用性高:服务可以被多个系统重用,提高了系统的可重用性。
- 可维护性好:服务之间通过接口进行通信,降低了系统间的耦合度,提高了系统的可维护性。
- 灵活性强:服务可以独立部署、运行和扩展,提高了系统的灵活性。

SOA 架构存在以下的问题:
- 服务接口定义复杂:服务接口需要定义清晰、明确的规范,以确保不同系统之间的通信。
- 服务治理困难:随着服务数量的增加,服务治理和监控变得越来越复杂。
- 技术门槛高:SOA 架构需要采用一些特定的技术框架和工具来支持服务注册、自动发现、服务调用等功能。

SOA架构强调服务的集成,它通过企业服务总线(Enterprise Service Bus,ESB)来管理所有业务的执行流程。基于各基础服务,可以将业务过程用类似BPEL(业务流程执行语言)流程的方式进行编排。架构如图2所示。

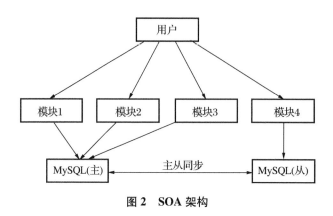

图 2　SOA 架构

2.2.4　微服务架构

现在流行的微服务架构基于SOA的思想,但是不再强调企业服务总线,而采用了去中心化的思想,通过网关对服务进行编排。为了满足现在越来越大的需求,对系统进行了更细粒度的划分,有利于资源的重复利用,拆分后的每个服务都只完成属于自己的某个特定功能,使服务之间的耦合度大大降低。但是划分后服务过多,会导致治理成本变高,不利于运维人员进行系统维护。

服务架构的优点包括：

- 独立性强：每个微服务都独立运行、独立部署和扩展,降低了系统间的耦合度。
- 技术栈灵活：不同微服务可以采用不同的技术栈和语言进行开发。
- 故障隔离性好：某个微服务的故障不会影响其他微服务的正常运行。

微服务架构也存在一些不足：

- 复杂性增加：随着微服务数量的增加,系统的复杂性也随之增加,需要更多的开发、测试、运维人员参与。
- 服务间通信开销大：微服务之间的通信需要通过网络进行,增加了通信开销和延迟。
- 服务治理复杂：需要对大量的微服务进行注册、发现、调用、监控、治理等操作,复杂度较高。

微服务架构如图3所示。

2.2.5　服务网格

针对微服务架构面临的问题,人们发现了更好的解决方案,OSI(开放系统互连参考模

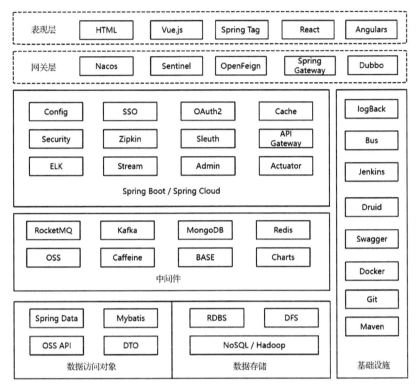

图 3 微服务架构

型)定义了开放系统的层次结构、层次之间的相互关系以及各层所包括的可能的任务,上层并不需要对底层具体功能有详细的了解,只需按照定义的准则协调工作即可。因此,我们也可参照 OSI 七层模型将公用库设计为位于网络栈和应用业务逻辑之间的独立层,即透明网络代理,新的独立层完全从业务逻辑中抽离,作为独立的运行单元,与业务不再直接紧密关联。通过在独立层的透明网络代理上实现负载均衡、服务发现、熔断、运行时动态路由等功能,该透明代理在业界有一个非常新颖的名字:Service Mesh(服务网格)。

在这种方案中,Service Mesh 作为独立运行层,它很好地解决了微服务所面临的挑战,使应用具备处理网络弹性逻辑和提供可靠交互请求的能力。它使得耦合性更低、灵活性更强,跟现有环境的集成时间和人力代价更小,也提供多语言支持、多协议支持,运维和管理成本更低。最主要的是开发人员只需关注业务代码逻辑,而不需要关注业务代码以外的其他功能,即 Service Mesh 对开发人员是透明的。

针对资源分享应用后台管理系统的特点,我们可以从以下几个方面考虑选择合适的系统架构:

- 业务需求:根据业务需求,选择能够满足系统功能性需求的架构。例如,如果系统需要支持高并发、大数据量等场景,可以考虑采用分布式架构或微服务架构。
- 技术团队实力:评估技术团队的技术实力和经验,选择适合团队技术栈的架构。例如,如果团队对 Java 技术栈比较熟悉,可以选择基于 Java 的微服务框架进行开发。

- **成本考虑**：考虑系统的开发、维护、升级等成本，选择性价比高的架构。例如，对于初创公司或小型项目，可以选择单体架构或简单的分布式架构来降低成本。
- **可扩展性和可靠性**：考虑系统的可扩展性和可靠性要求，选择能够支持系统平滑扩展和容错能力强的架构。例如，微服务架构通过服务拆分和独立部署，可以实现系统的水平扩展和容错能力提升。

综上所述，资源分享应用后台管理系统的架构选择需要根据业务需求、技术团队实力、成本考虑以及可扩展性和可靠性要求等多方面因素进行综合评估。在实际项目中，可以根据具体情况选择合适的架构或多种架构组合使用，以满足系统的各项需求。

3 项目技术选型

3.1 任务描述

为了确保项目的顺利进行，波哥特别强调了技术选型的重要性。技术选型不仅关系到项目的开发效率、运行性能和维护成本，更是项目成功与否的关键因素之一。在技术选型过程中，波哥和团队成员们需要深入探讨以下内容：

1. 需要分析项目的具体需求，明确项目目标，并据此来评估所需技术的适用范围和可行性。
2. 需要评估市场上的各种技术，包括技术的成熟度、稳定性、性能以及社区支持等方面，确保所选技术能够满足项目的实际需求。
3. 需要结合团队的技术能力和经验，选择最适合团队的技术栈，以提高开发效率和质量。

在技术选型中，后端框架、数据库、缓存技术和前端框架等都是需要重点考虑的因素。这些技术的选择将直接影响项目的整体性能和用户体验。因此，团队需要仔细分析每种技术的优缺点，并结合项目的实际情况进行选择。除了技术选型本身，团队还需要考虑技术的可扩展性和未来发展趋势。随着项目的不断发展和市场的不断变化，所选技术需要能够适应并满足项目的未来发展需求。因此，团队需要在技术选型过程中充分考虑这些因素，确保所选技术具有足够的可扩展性和灵活性。

波哥鼓励团队成员们积极参与技术选型工作，并强调要充分考虑各种因素，做出明智的选择。他相信通过团队的共同努力和细致的分析评估，一定能够选择出最适合项目的技术，为项目的成功奠定坚实的基础。

3.2 任务分析

3.2.1 软件开发模式的演变过程

在 Web 开发的历史长河中，技术的演进与变革总是与解决特定问题和满足日益增长的需求息息相关。从最初的简单页面展示,到现如今复杂的分布式系统架构,每一步的发展都标志着 Web 开发模式的重大突破。特别是在资源分享应用后台管理系统的开发过程中,技术框架的演变尤为显著。以下将详细分析这些技术框架的演变过程,以及它们如何逐步推动 Web 开发向前发展。

1. 早期阶段:Servlet(服务连接器)与 JSP(动态网页技术标准)

在 Java Web 开发的早期,Servlet 和 JSP 技术占据了主导地位。Servlet 作为 Java Web 应用程序的基础组件,负责处理客户端请求并生成响应。它可以直接与 Web 服务器进行交互,接收客户端的请求数据,并将处理结果返回给客户端。然而,Servlet 在处理业务逻辑和生成动态页面时存在较大的局限性,因为它需要将业务逻辑和页面展示混合在一起,导致代码难以维护和扩展。

为了解决 Servlet 的局限性,JSP 技术应运而生。JSP 允许开发者在 HTML 页面中嵌入 Java 代码片段,通过 JSP 标签库和表达式语言等特性,将 Java 代码与 HTML 页面分离,实现业务逻辑和页面展示的分离。JSP 页面在服务器上被编译成 Servlet,然后执行并生成 HTML 响应返回给客户端。这种方式在一定程度上提高了代码的可维护性和可扩展性,但仍然存在一定的局限性。

首先,JSP 页面中的 Java 代码片段难以管理和维护,因为它们分散在多个页面中,难以进行统一的管理和修改。其次,JSP 页面中的 Java 代码与 HTML 代码混合在一起,不利于代码的重用和组件化开发。最后,JSP 页面在编译时需要加载大量的类库和依赖项,导致启动时间较长,影响系统性能。

2. MVC(Model-View-Controller)架构的兴起与 Struts

为了解决 Servlet 和 JSP 带来的问题,MVC 架构开始被广泛应用。MVC 架构将应用程序分为三个核心部分:模型(Model)、视图(View)和控制器(Controller)。模型负责处理业务逻辑和数据存储,视图负责展示用户界面,控制器负责接收用户请求并调用相应的模型进行处理,然后将处理结果传递给视图进行展示。

MVC 架构的优点在于实现了业务逻辑、用户界面和用户输入处理的分离,使得代码更加清晰、易于管理和维护。同时,MVC 架构还提高了代码的可重用性和可扩展性,降低了系统的耦合度。

在 MVC 架构的实践中,Apache Struts 是最早实现 MVC 设计模式的 Web 框架之一。Struts 通过其控制器(Action Servlet)来处理请求,并通过配置文件(struts-config.xml)管理路由和组件的映射关系。Struts 框架提供了丰富的标签库和验证机制,使得开发者能够更快速地构建 Web 应用程序。然而,Struts 框架也存在一些局限性:首先,它的配置相对烦琐,需要编写大量的 XML 配置文件来定义路由和组件的映射关系。其次,Struts 框架在处理大量并发请求时性能较差,容易成为系统的瓶颈。

3. Spring 框架的崛起

为了解决企业级应用开发中的复杂性和烦琐性,Spring 框架应运而生。Spring 是一个轻量级的Java EE应用程序框架,它提供了全面的编程和配置模型,旨在简化企业级应用的开发和维护。Spring的核心功能包括依赖注入(DI,Dependency Injection)、面向切面编程(AOP,Aspect Oriented Programming)、事务管理、安全性等。

在Web开发领域,Spring框架提供了Spring MVC模块来支持MVC架构的实现。Spring MVC基于Java实现,采用了请求驱动的设计模式,支持灵活的URL映射、视图解析和参数绑定等功能。与Struts相比,Spring MVC具有更简洁的配置和更灵活的控制器方法。同时,Spring MVC还支持RESTful Web服务的开发,并提供了丰富的扩展点和插件机制,使得开发者能够根据自己的需求进行定制和扩展。

除了Spring MVC之外,Spring框架还提供了其他许多实用的功能。例如,Spring的依赖注入功能可以帮助开发者实现组件之间的松耦合;Spring的AOP功能可以帮助开发者实现跨切面的功能扩展;Spring的事务管理功能可以帮助开发者简化数据库操作的复杂性。这些功能使得Spring框架成为企业级应用开发的首选框架之一。

4. Spring MVC 的流行

随着Spring框架的不断发展和完善,Spring MVC作为MVC架构在Spring生态系统中的实现逐渐被广泛接受和应用。Spring MVC不仅继承了MVC架构的优点,还结合了Spring框架的强大功能,为开发者提供了更加灵活、可扩展和可维护的Web开发解决方案。

首先,Spring MVC支持基于注解的配置方式,使得开发者能够更加简洁地定义路由和组件的映射关系。与Struts相比,Spring MVC的配置更加直观和易于理解。其次,Spring MVC提供了丰富的控制器方法注解和参数绑定机制,使得开发者能够更快速地构建RESTful Web服务和处理各种请求类型。此外,Spring MVC还支持多种视图技术(如JSP、Thymeleaf等),使得开发者能够根据自己的需求选择最适合的视图技术进行页面展示。

除了以上优点之外,Spring MVC还具有以下特点,进一步推动了其在Web开发领域的流行:

- 高度集成与可扩展性

Spring MVC作为Spring框架的一部分,与Spring的其他模块(如Spring Security、Spring Data等)无缝集成,使得开发者能够轻松构建功能强大的Web应用程序。同时,Spring MVC的扩展性也非常好,开发者可以通过自定义拦截器、消息转换器、视图解析器等来扩展系统的功能。

- 强大的异常处理机制

在Web开发中,异常处理是不可避免的一部分。Spring MVC提供了强大的异常处理机制,允许开发者通过注解或配置的方式定义全局或局部的异常处理器。这些异常处理器能够捕获并处理请求过程中发生的异常,保证应用程序的健壮性和用户体验。

- 灵活的国际化支持

对于需要支持多语言环境的Web应用程序,Spring MVC提供了灵活的国际化支持。通过定义资源文件和配置相应的消息源,Spring MVC能够根据用户的语言环境显示不同的页面内容。这使得开发者能够轻松地构建多语言版本的 Web 应用程序。

- 丰富的社区支持和文档

Spring MVC 作为一个广泛使用的 Web 框架,拥有庞大的社区支持和丰富的文档资源。这些文档和资源为开发者提供了大量的示例代码、最佳实践、常见问题解答等,帮助开发者更快地掌握 Spring MVC 的使用技巧并解决开发过程中遇到的问题。

5. Spring Boot 的简化配置

随着微服务架构的兴起和云计算的普及,对于快速构建和部署 Web 应用程序的需求越来越高。为了满足这一需求,Spring Boot 应运而生。Spring Boot 是一个用于简化 Spring 应用的初始搭建以及开发过程的框架。它通过约定大于配置的方式,减少了大量烦琐的配置工作,使得开发者能够更快速地构建和部署 Web 应用程序。

在 Spring Boot 中,开发者只需要添加必要的依赖项并在主类上添加@SpringBootApplication 注解,就可以快速地启动一个 Web 应用程序。Spring Boot 还提供了自动配置功能,能够根据项目的依赖项自动配置相关的组件和参数,进一步简化了开发过程。

从早期的 Servlet 和 JSP 技术到 MVC 架构的兴起与 Struts 的应用,再到 Spring 框架的崛起、Spring MVC 的流行以及 Spring Boot 的简化配置,这些技术框架的发展为 Web 开发提供了更加高效、灵活和可维护的解决方案。每一次技术的迭代都是为了更好地适应开发者的需求,提高开发效率,同时也为了更好地适应市场和技术环境的变化。每种技术的引入都是对前者的改进或补充,让开发者可以根据项目需求和个人偏好选择最适合的工具和框架。

3.2.2 开源项目 Geeker-Admin

Geeker-Admin——一款基于 Vue 3.4、TypeScript、Vite5、Pinia、Element Plus 开源的后台管理框架,使用目前最新技术栈开发。项目提供强大的 ProTable 组件,可在一定程度上提高开发效率。另外 Geeker-Admin 还封装了一些常用组件、Hooks 函数、指令、动态路由、按钮级别权限控制等功能。

Vue 是一套用于构建用户界面的渐进式框架。与其他大型框架不同的是,Vue 被设计为可以自底向上逐层应用。Vue 只关注视图层,不仅易于上手,还便于与第三方库或既有项目整合。Vue 提供了更加简洁、更易于理解的 API(Application Programming Interface,应用程序编程接口),与其他库的结合也是很方便,便于项目的整合,使得我们能够快速地上手并使用 Vue。同时,其还可以进行组件化开发,使代码编写量大大减少,更加易于理解。Vue 最突出的优势在于可以对数据进行双向绑定,使用 Vue 写出来的界面效果本身就是响应式的,这使网页在各种设备上都能显示出非常好看的效果,相比传统的页面通过超链接实现页面的切换和跳转,在 Vue 中使用路由,不会引起页面刷新。

TypeScript 是由微软开发的开源、跨平台网络编程语言,是 JavaScript 的超集。其本质是向该语言添加了可选的静态类型和基于类的面向对象编程,通过扩展 JavaScript 的语法,使得现有的 JavaScript 程序均可在 TypeScript 环境中运行。TypeScript 也可为已存在的

JavaScript 库添加类型信息的头文件，扩展其对于流行库的支持。TypeScript 必将是未来 Web 技术主流开发语言。

Vite 是一种新型前端构建工具，能够显著提升前端开发体验。它主要由两部分组成：一个开发服务器，它基于原生 ES 模块提供了丰富的内建功能，如速度快到惊人的模块热更新（HMR）；一套构建指令，它使用 Rollup 打包代码，并且是预配置的，可输出用于生产环境的高度优化过的静态资源。Vite 意在提供开箱即用的配置，同时它的插件 API 和 JavaScript API 带来了高度的可扩展性，并有完整的类型支持。

对于 Vite 而言，它的一个最大的核心优势就是快，特别是在大型项目中，它在开发时的构建速度基本都可以维持在 1 s 以内，这种优势是 Webpack 所不具备的。Vite 提供依赖预构建的功能，其中一个非常重要的目的就是为了解决 CommonJS 和 UMD 兼容性问题。目前 Vite 会先将 CommonJS 或 UMD 发布的依赖项转换为 ESM 之后，再重新进行编译。

Pinia 是一个用于 Vue 的状态管理库。这个库提供了一个简单且直观的 API 来管理应用程序的状态。Pinia 是作为 Vue 官方状态管理库 Vuex 的一种轻量级替代方案而创建。它的目标是提供一个更加简单和灵活的 API，同时还保留 Vuex 的主要功能。

Pinia 的 API 设计简洁明了，易于理解和使用。这使得状态管理变得更加直观，让开发者可以更快速地上手和使用。Pinia 集成了 Vuex 的开发者工具，使得开发者可以轻松地跟踪状态变化和调试代码。与 Vuex 中的模块不同，Pinia 允许开发者自由地组织和管理状态。这使得在大型应用中管理状态变得更加灵活和方便。Pinia 能够与 Vue 3 的组合式 API 无缝集成，使得状态管理代码更加的清晰和模块化。Pinia 提供了出色的 TypeScript 支持，允许开发者在编写代码时获得类型提示和自动补全，提高代码的质量和可维护性。

Element Plus 是一个基于 Vue 3 的高质量 UI 组件库。它为开发者提供了一套丰富的 UI 组件和扩展功能，帮助开发者快速构建高质量的 Web 应用。其包含丰富的组件和扩展功能，例如表格、表单、按钮、导航、通知等，让开发者能够快速构建高质量的 Web 应用。Element Plus 的设计理念是：提供开箱即用的 UI 组件和扩展功能，帮助开发者快速构建应用程序，同时提供详细的文档和教程，让开发者更好地掌握和使用 Element Plus。

4 项目开发规划

4.1 任务描述

波哥深知项目开发规划的重要性。他召集团队成员，并提出了一系列关键的开发阶段，包括技术选型、系统设计、编码实现、测试验收等，并强调了每个阶段的任务目标。团队成员们对项目的整体开发规划有了更清晰的认识，并对未来的工作充满了期待和信心。

4.2 任务分析

本项目将按照以下步骤进行开发：

- 任务一:设计数据库。分析业务需求,定义数据模型和关系;选择合适的数据库系统;设计数据表及其关系;确保数据的完整性和安全性。
- 任务二:设计接口文档。确定各功能的接口需求;撰写详细的 API 文档,包括请求类型、参数、响应格式等;确保文档的清晰性和易用性。
- 任务三:搭建后台管理系统。选择合适的开发框架和技术栈;设置项目结构;配置基本的服务,如服务器通信、数据库连接等。
- 任务四:用户管理模块开发。设计用户认证和授权流程;实现注册和登录接口;开发用户角色和权限控制功能。
- 任务五:积分管理模块开发。定义积分规则;实现积分增减的逻辑;提供积分查询接口。
- 任务六:通知管理模块开发。设计通知模板;实现通知的创建、发送和管理功能;确保通知的及时性和准确性。
- 任务七:标签管理模块开发。实现标签的创建、编辑、删除和查询功能;保证标签数据的一致性。
- 任务八:分类管理模块开发。设计分类体系;实现分类的添加、编辑、删除和查询功能。
- 任务九:资源管理模块开发。这是核心模块,实现资源的审核、删除等核心功能。
- 任务十:首页仪表盘开发。展示关键数据指标,如用户数量、资源数量、访问量等。
- 任务十一:后台管理系统打包部署。完成系统的测试和优化;打包应用;配置服务器和部署应用;监控系统运行状态,确保无误,并准备上线。

本项目的整体开发规划如下:

细分任务
准备任务:了解项目
任务一:设计数据库
任务二:设计接口文档
任务三:搭建后台管理系统
任务四:用户管理模块开发
任务五:积分管理模块开发
任务六:通知管理模块开发
任务七:标签管理模块开发
任务八:分类管理模块开发
任务九:资源管理模块开发
任务十:首页仪表盘开发
任务十一:后台管理系统打包部署

5 任务总结

本次任务我们主要完成了以下三个关键任务,为资源分享应用后台管理系统的开发工作奠定了坚实的基础。

1. 初识项目:我们首先对即将开发的资源分享应用后台管理系统进行了全面的了解,明确了项目的需求、目标以及整体架构。通过这一步骤,我们对项目有了清晰的认知,为后续的开发工作做好了准备。

2. 项目架构设计:在了解项目的基础上,我们进行了项目架构设计。通过综合考虑项目的需求、团队能力、市场前景等因素,我们选择了最适合的架构模式,并制定了详细的架构设计方案。这为我们后续的开发工作提供了明确的指导。

3. 项目开发规划:我们制定了详细的项目开发规划。从准备任务到各个功能模块的开发,再到最终的打包部署,我们制定了清晰的任务分配。这一规划将确保我们的开发工作能有条不紊地进行,提高开发效率和质量。

通过完成以上三个任务,我们已经为资源分享应用后台管理系统的开发工作做好了充分的准备。接下来,我们将按照规划逐步推进项目的开发,确保项目能够按时、高质量地完成。

扫描二维码,参考文件夹中"了解项目-SOA架构图""了解项目-单体架构图"和"了解项目-微服务架构图"三个文件。

任务一

设计数据库

波哥、小南和小工等人组成的项目团队已经深入了解了项目的需求和整体架构。现在,他们即将进入数据库设计的关键阶段。

波哥召集了团队成员:"各位,我们接下来要进入数据库设计阶段了。这是整个项目中非常重要的一环,因为它直接关系到我们系统的稳定性和可扩展性。"

◇ 任务点

- 设计数据库表;
- 分析表结构;
- 编写数据库设计文档。

◇ 任务计划

- 任务内容:设计项目的数据库;
- 任务耗时:预计完成时间为 30 min～1 h;
- 任务难点:合理的数据库设计技巧。

1 数据库设计

1.1 任务描述

波哥与小南和小工等团队成员开始共同讨论并规划数据库设计的关键步骤。

首先,波哥强调了数据库设计的核心地位,指出了解当前主流数据库系统是第一步。他建议团队成员深入研究不同数据库的特点和优势,以便根据项目需求选择最合适的数据库。

在讨论中,小南提出了一个关键问题:"我们如何确保数据库设计既高效又易于维护?"

波哥回应道:"首先,我们需要从业务需求出发,明确需要存储哪些数据以及这些数据之间的关系。其次,在设计数据库表结构时,我们要注重数据的存储效率,但同样重要的是,我们还需要考虑数据的可读性和可维护性。这样,在未来的开发和维护过程中,我们才能更加便捷地管理和更新数据。"

小工则关注到安全性问题:"我们如何确保数据库的安全性呢?"

波哥进一步解释道:"RBAC(Role-Based Access Control,基于角色的访问控制)权限管理功能是关键。在设计数据库时,我们要充分考虑如何实现这一功能。通过为不同的用户角色分配不同的权限,我们可以控制用户对数据的访问和操作,从而保护业务数据的安全性和完整性。"

在波哥的指导下,小南和小工积极参与了数据库设计的讨论和规划工作。他们结合所学的数据库知识,提出了多种设计方案,并与波哥进行了深入的讨论和比较。经过几轮讨论和修改,团队最终确定了一个既满足业务需求又具有良好可读性和可维护性的数据库设计方案。

通过这次数据库设计的讨论和规划,不仅提高了小南和小工的数据库设计能力,也让他们更加深入地理解了业务需求与数据库设计之间的关系。

1.2 任务分析

1.2.1 关系型数据库

关系型数据库建立在关系型数据模型的基础上,是借助于集合代数等数学概念和方法来处理数据的数据库。这种数据库的数据存储以行和列的形式展开,其基本单位为表。在关系型数据库中,实体及实体间的关系均通过一种单一的结构类型表示,即二维表。多个这样的表通过关系和约束相互联系,共同构成一个关系型数据库。用户通过结构化查询语言(SQL)来执行增加、删除、修改和查询数据库中数据的操作。

目前市面上使用较多的关系型数据库包括甲骨文公司的 Oracle、微软公司的 SQL Server、IBM 公司的 DB2、Sybase 公司的 Sybase、英孚美软件公司的 Informix,以及开源的 MySQL 和 PostgreSQL 等。其中,Oracle 数据库常用于大型企业中的关键业务系统,因其高性能、高可靠性和强大的数据处理能力而受到青睐。SQL Server 广泛应用于各种企业级应用,特别是与其他微软技术集成的环境中。MySQL 由于其开源性、免费以及出色的性能,成为许多开发者和中小企业的首选。这些数据库各有其特点和优势,适用于不同的应用场景和需求。

1. Oracle

Oracle Database,又名 Oracle RDBMS,或简称 Oracle,是甲骨文公司的一款关系数据库管理系统。它在数据库领域一直处于领先地位。可以说 Oracle 数据库系统是世界上流行的关系数据库管理系统,系统可移植性好、使用方便、功能强,适用于各类大、中、小微机环境。它是一种高效率、可靠性好的、适应高吞吐量的数据库方案。

- 特点:

支持复杂的事务处理。

提供全面的数据加密和安全管理。

强大的 PL/SQL 程序设计语言。

高可扩展性和可靠性。
- 应用场景：

大型企业和金融机构的数据存储和处理。

高端商务应用系统。

数据仓库和在线事务处理（OLTP）。

2. MySQL

MySQL是一种关系型数据库管理系统，运用SQL语句来进行数据库管理和数据查询工作，支持多种操作系统开发。从执行原理上讲，数据库是通过计算机整理后形成的数据，可存储在一个或多个文件中，而管理这个数据库的工具便是数据库管理系统。MySQL可以存储大量的数据并支持高并发处理，具备高性能和可靠性，是开发常用的数据库之一。

- 特点：

易于安装和使用。

高性能，尤其在读取操作中表现突出。

可扩展性强，支持大量的并发连接。

丰富的社区资源和丰富的第三方应用支持。

- 应用场景：

网站后端数据库。

应用程序快速开发。

轻量级到中等规模的数据库解决方案。

3. PostgreSQL

PostgreSQL是一种开源的关系型数据库管理系统。它具有强大的功能和广泛的应用场景。PostgreSQL以其稳定性和可靠性而闻名，被广泛使用。它提供了丰富的SQL功能，支持复杂的数据结构和查询操作。它还具有良好的可扩展性，可以通过插件和扩展来增强其功能。PostgreSQL社区活跃，拥有丰富的文档和资源。无论是小型项目还是大型企业应用，PostgreSQL都是一个受欢迎的选择，它为用户提供了高效、可靠的数据管理解决方案。

- 特点：

支持高级的数据类型和函数。

可扩展的架构，用户可以定义自己的数据类型、索引方法、功能等。

高度兼容SQL标准。

强大的并发处理能力。

- 应用场景：

企业级应用系统。

地理信息系统（GIS）。

数据分析和数据仓库。

以上三种关系型数据库特性对比见表1。

表1 三种关系型数据库特性对比

特性	Oracle	MySQL	PostgreSQL
开发商	Oracle 公司	Oracle 公司	开源社区
数据安全性	高级安全功能	中等,提供基本安全功能	高,提供全面的安全功能
事务支持	强大的事务支持,完整的 ACID 兼容	支持事务,ACID 兼容	强大的事务支持,完整的 ACID 兼容
扩展性	高,适合大型企业级应用	中等,适合中小型应用	高,适合需要复杂查询的大型应用
性能	高性能,尤其是在复杂查询上	高性能,尤其是读取速度	高性能,尤其是在复杂查询上
程序设计语言	PL/SQL	SQL	SQL,支持存储过程和函数
主要应用场景	大型企业、金融机构	网站后端、轻量级到中等规模数据库	企业级应用、数据分析、GIS
使用难度	高,需要专业知识和经验	中等,相对容易上手	中等,功能丰富但学习曲线适中
成本	高,需要购买许可证及可能的额外支持成本	低(社区版免费),企业版需要付费	低,完全免费

1.2.2 非关系型数据库

　　非关系型数据库,通称为 NoSQL(Not Only SQL),这个术语强调这类数据库不单单依赖于传统的结构化查询语言 SQL。区别于关系型数据库的严格表格结构,NoSQL 数据库采用更灵活的数据模型,如键值对、列存储、文档和图形等。这种多样的数据模型设计使得 NoSQL 数据库特别适合处理结构松散或迅速变化的大规模数据集,以及高并发的数据访问请求。

　　尽管 NoSQL 数据库通常不提供完整的 ACID 事务特性,许多现代 NoSQL 系统通过不同的设计和技术策略提供了类似的数据一致性和持久性保障。由于其开源和免费的特性,NoSQL 数据库在互联网和企业应用中得到了广泛应用。例如,MongoDB 作为一款文档型数据库,优于处理复杂数据结构和动态模式,广泛应用于 Web 应用和内容管理系统。Redis 则以其出色的读写性能广泛应用于数据缓存和会话管理。Cassandra 和 HBase 等列存储数据库在处理大规模分布式数据时显示出其优势,适合用于需要高并发和高可用性的应用场景。

　　此外,随着技术的发展,其他专用的 NoSQL 数据库也在特定领域显示出其独特的价值。例如,Neo4j 这类图形数据库非常适合处理复杂的网络结构数据,而 Elasticsearch 作为一款强大的搜索引擎,擅长处理大规模文档的搜索和分析任务。这些数据库各有特点,用户可以根据具体的应用需求选择最适合的 NoSQL 解决方案。

表2 关系型数据库与非关系型数据库对比

特性	关系型数据库	非关系型数据库
数据模型	表格模型,数据以行和列的形式存储	键值对、文档、列存储、图形等多种数据模型
查询语言	使用SQL(结构化查询语言),具有强大的查询能力	通常使用自定义查询语言,简单但可能不如SQL强大
事务支持	支持完整的ACID事务特性,确保数据的一致性和可靠性	通常不支持全ACID特性,某些支持最终一致性
扩展性	垂直扩展,通常通过增加服务器的处理能力进行扩展	水平扩展,通过增加更多服务器来处理更大的数据量
应用场景	复杂查询、事务完整的场景,如银行、ERP系统	大数据处理、实时应用、灵活的数据存储,如社交网络、实时分析
典型代表	Oracle、MySQL、PostgreSQL	MongoDB、Redis、Cassandra、HBase

在关系型数据库与非关系型数据库之间进行对比后(表2),选择MySQL的原因显得尤为多样化且综合。MySQL作为一个广泛应用的数据库管理系统,以其深厚的用户基础和活跃的社区支持著称。这种庞大的支持网络意味着开发者可以获得丰富的资源和高效的问题解决途径。它的性能在处理中小型项目时表现出色,并且具有良好的可扩展性,能够满足从小型企业到大型企业不同规模的需求。MySQL的开源特性进一步增加了其成本效益,使之成为预算有限的项目的理想选择。

自1995年以来,MySQL已被广泛应用于各种行业和应用中,这种广泛的使用证明了其稳定性和可靠性。作为开源产品,MySQL背后有一个非常活跃的社区,为用户提供大量的文档、教程和支持资源,此外还有众多第三方工具和用户界面支持MySQL,极大地方便了数据库的管理和优化。其高性能的数据处理能力可以通过各种优化策略进一步提高,例如索引、查询优化和配置调整,确保了处理效率和速度。

MySQL支持多种存储引擎,包括InnoDB(提供事务支持)和MyISAM(优化了读操作的性能),这使得用户可以根据具体需求选择最适合的存储引擎,以优化应用的性能和数据完整性。这种灵活性使得MySQL可以应用于多种不同的技术场景,从Web应用到复杂的数据仓库系统。

综合来看,MySQL的成熟与普及、社区与支持、卓越性能、存储引擎的灵活性和卓越的成本效益,使得它在关系型数据库中脱颖而出,成为许多组织和开发者的首选。这些因素共同构成了选择MySQL的有力理由,尤其是在需要一个可靠、高效且经济的数据管理平台时。

2 数据表结构设计

2.1 任务描述

数据库表结构设计对于整个系统的稳定性和性能有着至关重要的影响。

波哥告诉小南和小工,设计数据表结构不仅仅是为了存储数据,更重要的是如何合理、有效地组织和存储这些数据。明确业务需求并确定所需存储的数据及其相互关系是设计数据表结构的起点。同时,他详细解释了如何根据这些数据的特点和访问模式来设计数据表,包括选择适当的字段、数据类型以及设计表与表之间的关联关系。

一个优秀的数据表结构设计能够显著提升系统的性能。例如,合理的索引设计可以加速数据查询的速度,而适当的数据冗余则可以在一定程度上提高数据的可用性和容错性。

在讨论中,波哥还提到了设计数据表结构时需要遵循的一些原则。比如:简洁性,即避免不必要的复杂性和冗余;规范性,即遵循统一的命名规则和字段类型选择;可扩展性,即考虑到系统未来的发展需求,为可能的扩展预留空间;安全性,即确保数据的安全性和完整性。

经过这次深入的讨论,小南和小工对数据表结构设计的重要性和关键要素有了更深入的理解。他们认识到,在设计过程中需要综合考虑业务需求、系统性能、安全性和可扩展性等多个方面,以确保最终设计出的数据表结构既满足业务需求,又具有良好的性能和安全性。接下来,他们将根据这些原则和要点,开始着手设计资源分享应用后台管理系统的数据表结构。

2.2 任务分析

在开发过程中,通常使用实体对象来操作数据,而存储数据时,需要把实体对象中的数据存放到数据表中,也就是说实体对象中的每个属性对应数据表中的一个字段。因此可以根据实体对象来设计数据表结构。

根据项目功能分析,本系统的实体对象包括用户实体、用户行为实体、积分日志实体、资源分类实体、资源实体、标签实体、公告实体、系统字典实体、系统字典配置实体、系统用户管理实体、角色管理实体、用户角色关系实体、系统菜单实体、角色菜单关系实体等。接下来根据上述实体对象设计数据表结构,具体如下。

2.2.1 用户表 t_user

用户表 t_user 通常是核心部分之一,因为它存储了与系统交互的实体的基本信息。该表是系统用户认证、权限分配、个性化设置等功能的基础。用户表的主要功能有:

- 用户认证:存储用户登录所需的必要信息(如用户名、密码等),用于验证用户的身份。

- 用户信息管理：存储用户的个人信息（如姓名、邮箱、手机号等），以便进行用户联系、信息推送等操作。
- 用户状态管理：通过状态字段记录用户的账户状态（如活跃、禁用、删除等），以便进行用户权限控制和统计分析。
- 关联其他业务数据：作为业务数据表的外键，与用户的其他操作（如订单、评论、反馈等）进行关联，构建完整的用户行为数据链。

用户表 t_user 的表结构如表 3 所示。

表 3 用户表的设计

字段名称	字段类型	约束	为空	字段含义
avatar	varchar(1000)		NO	头像
birthday	varchar(255)		YES	生日
bonus	int		YES	积分
create_time	timestamp		NO	创建时间
delete_flag	tinyint		NO	删除标识(0-未删除,1-删除)
enabled	tinyint		NO	是否可用(0-冻结,1-可用)
gender	tinyint		YES	性别(0-男,1-女)
nickname	varchar(36)		NO	昵称
phone	varchar(11)		YES	手机号
pk_id	int	主键	NO	主键
remark	varchar(255)		YES	备用字段
update_time	timestamp		NO	更新时间
wx_open_id	varchar(255)		YES	微信 openId

2.2.2 用户行为表 t_action

用户行为表 t_action 是用于记录系统中用户行为的数据表。通过分析这些用户行为数据，企业可以了解用户的活跃程度、使用习惯以及潜在需求，从而优化产品设计、改进用户体验，并提升业务效率。用户行为表的主要功能有：

- 行为分析：收集并存储用户的行为数据，用于后续的数据分析和挖掘，以揭示用户的使用习惯、偏好和需求。
- 用户画像：基于用户行为数据构建用户画像，帮助企业更准确地了解用户群体，为个性化推荐、精准营销等提供支持。
- 产品优化：根据用户行为数据发现产品中存在的问题和不足之处，从而进行针对性的优化和改进。
- 安全监控：监控和记录异常的用户行为，以便及时发现和处理安全漏洞或违规行为。

用户行为表 t_action 的表结构如表 4 所示。

表 4　用户行为表的设计

字段名称	字段类型	约束	为空	字段含义
create_time	timestamp		NO	创建时间
delete_flag	tinyint		NO	删除标识(0-未删除,1-删除)
pk_id	int	主键	NO	主键
resource_id	int		NO	资源 id
type	tinyint		NO	类型(0-收藏,1-发布,2-兑换,3-点赞)
update_time	timestamp		NO	更新时间
user_id	int		NO	用户 id

2.2.3　积分日志表 t_bonus_log

积分日志表 t_bonus_log 是用于记录系统中用户积分变化情况的日志数据表。通过这张表，企业可以追踪用户积分的增减历史，分析用户积分行为，并为后续的积分运营策略提供数据支持。t_bonus_log 表的主要功能有：

- 积分变化记录：详细记录每次用户积分增减的详细信息，包括积分数额、操作类型、时间戳等。
- 用户积分分析：基于日志数据，分析用户积分的获取途径、消费习惯以及积分余额的分布情况。
- 积分策略优化：通过用户积分行为的分析，发现当前积分策略存在的问题，并进行相应的优化和调整。
- 积分异常监控：监控异常的积分增减行为，如大量积分异常增加或减少，以便及时发现并处理潜在的安全问题或系统错误。

积分日志表 t_bonus_log 的表结构如表 5 所示。

表 5　积分日志表的设计

字段名称	字段类型	约束	为空	字段含义
bonus	int		NO	积分
content	varchar(255)		NO	用户行为
create_time	timestamp		NO	创建时间
delete_flag	tinyint		NO	删除标识(0-未删除,1-删除)
pk_id	int	主键	NO	主键
update_time	timestamp		NO	更新时间
user_id	int		NO	用户 id

2.2.4 资源分类表 t_category

资源分类表 t_category 是用于存储系统中资源分类信息的数据表。通过这张表,系统可以对资源进行有序的分类管理,方便用户查找和浏览。资源分类表 t_category 的主要功能有:

- 分类管理:提供对系统中各类资源的分类管理功能,如文章分类、商品分类、服务分类等。
- 层级结构:支持多级分类,形成树状结构,便于对资源进行更细粒度的划分。
- 关联资源:与其他资源表(如文章表、商品表等)建立关联,指定资源所属的分类。
- 查询优化:通过分类信息优化资源查询,提高检索效率。

资源分类表 t_category 的表结构如表 6 所示。

表 6 资源分类表的设计

字段名称	字段类型	约束	为空	字段含义
create_time	timestamp		NO	创建时间
delete_flag	tinyint		NO	删除标识(0-未删除,1-删除)
description	varchar(255)		YES	描述
pk_id	int	主键	NO	主键
title	varchar(255)		NO	分类名称
type	tinyint		NO	分类类型(0-网盘类型,1-资源类型)
update_time	timestamp		NO	更新时间

2.2.5 资源表 t_resource

资源表 t_resource 表是用于存储系统中各类资源详细信息的数据表。这些资源可以包括文章、图片、视频、文档等不同类型的内容。通过这张表,系统可以对资源进行统一的管理和查询。t_resource 表的主要功能有:

- 资源管理:提供对系统中各类资源的存储、修改、删除和查询等功能。
- 分类关联:与资源分类表 t_category 关联,指定资源所属的分类。
- 用户权限控制:根据用户的权限控制资源的访问和下载。
- 资源搜索:支持对资源的全文搜索和条件搜索,提高用户查找资源的效率。

资源表 t_resource 的表结构如表 7 所示。

表7 资源表的设计

字段名称	字段类型	约束	为空	字段含义
author	int		NO	发布人id
create_time	timestamp		NO	创建时间
delete_flag	tinyint		NO	删除标识(0-未删除,1-删除)
detail	text		NO	详情
disk_type	int		NO	网盘分类id
download_url	varchar(255)		NO	资源链接
is_top	tinyint		NO	是否置顶(0-否,1-是)
like_num	int		NO	点赞量
pk_id	int	主键	NO	主键
price	int		NO	价格
remark	varchar(255)		YES	审核结果描述
res_type	json		NO	资源分类id,多个
status	tinyint		NO	审核状态(0-待审核,1-通过,2-拒绝)
tags	json		NO	标签
title	varchar(255)		NO	标题
update_time	timestamp		NO	更新时间

2.2.6 标签表t_tag

标签表t_tag是用于存储系统中标签信息的数据表。标签通常用于对资源进行分类、标注或描述,以便于用户查找和识别相关内容。标签表t_tag的主要功能有:

- 标签管理:提供对系统中标签的添加、修改、删除和查询功能。
- 资源标注:与资源表(如t_resource)关联,用于标注资源所属的标签。
- 用户搜索:支持用户通过标签搜索相关资源,提高资源查找的效率和准确性。
- 推荐系统:作为推荐系统的一部分,根据用户的标签偏好推荐相关资源。

标签表t_tag的表结构如表8所示。

表8 标签表的设计

字段名称	字段类型	约束	为空	字段含义
create_time	timestamp		NO	创建时间
delete_flag	tinyint		NO	删除标识(0-未删除,1-删除)
description	varchar(255)		YES	描述

（续表）

字段名称	字段类型	约束	为空	字段含义
is_hot	tinyint		NO	是否热门(0-否,1-是)
pk_id	int	主键	NO	主键
title	varchar(255)		YES	标签名
update_time	timestamp		NO	更新时间

2.2.7 公告表 t_notice

公告表 t_notice 是用于存储系统中各类公告信息的数据表。这些通知可能是系统公告、用户提醒、活动通知等,用于向用户传达重要的信息或提醒。公告表 t_notice 的主要功能有:

- 公告管理:提供对系统中公告的添加、修改、删除和查询功能。
- 用户通知:将公告信息推送给用户,确保用户能够及时了解相关信息。
- 分类管理:支持对公告进行分类,方便用户按类别查看公告。
- 历史记录:保存公告的历史记录,方便用户查阅和回溯。

公告表 t_notice 的表结构如表9所示。

表9 公告表的设计

字段名称	字段类型	约束	为空	字段含义
pk_id	int	主键	NO	主键
title	varchar(255)		NO	标题
cover	varchar(255)		YES	封面图
content	varchar(255)		NO	内容
admin_id	int		NO	发布人 ID
is_top	tinyint		NO	是否置顶(0-否,1-是)
is_swiper	tinyint		NO	是否轮播(0-否,1-是)
delete_flag	tinyint		NO	删除标识(0-未删除,1-删除)
update_time	timestamp		NO	更新时间
create_time	timestamp		NO	创建时间

2.2.8 系统字典表 sys_dict

系统字典表 sys_dict 在数据库中主要用于存放多组值不变的基础数据,并进行统一的管理。系统字典表 sys_dict 的主要功能有:

- 存储基础数据:系统字典表主要用于存储那些在系统中不经常改变,但又需要被多次引用的基础数据。这些数据通常是业务中需要的静态信息,例如,一些固定的类别、状态、级别等。

- 统一管理:通过对这些基础数据的统一管理,可以避免在多个地方存储和维护相同的数据,提高数据的一致性和可维护性。
- 查询功能:系统字典表主要提供查询功能,支持通过编码、名称等字段查询对应的基础数据。这种查询功能通常用于在业务逻辑中根据特定的条件获取相应的数据。

系统字典表 sys_dict 的表结构如表 10 所示。

表 10 系统字典表的设计

字段名称	字段类型	约束	为空	字段含义
create_time	timestamp		NO	创建时间
delete_flag	tinyint		NO	删除标识(0-未删除,1-已删除)
description	varchar(255)		NO	描述
number	varchar(64)		NO	编号
pk_id	int	主键	NO	主键
title	varchar(255)		NO	名称
update_time	timestamp		NO	更新时间

2.2.9 系统字典配置表 sys_dict_config

系统字典配置表 sys_dict_config 通常用于存放与系统字典表 sys_dict 相关的配置信息。系统字典配置表 sys_dict_config 的主要功能有:

- 配置管理:sys_dict_config 表用于存储和管理与系统字典相关的配置信息。这些配置信息可能包括字典的显示顺序、是否启用、特定于某个业务场景的配置等。
- 与字典表的关联:该表通常与 sys_dict 表有关联,通过字典编码或其他标识符来引用具体的字典项。
- 业务逻辑控制:在某些业务逻辑中,需要根据 sys_dict_config 表中的配置信息来决定如何处理 sys_dict 表中的字典项。

系统字典配置表 sys_dict_config 的表结构如表 11 所示。

表 11 系统字典配置表的设计

字段名称	字段类型	约束	为空	字段含义
create_time	timestamp		NO	创建时间
delete_flag	tinyint		NO	删除标识(0-未删除,1-已删除)
dict_number	varchar(64)		NO	字典编号

（续表）

字段名称	字段类型	约束	为空	字段含义
pk_id	int	主键	NO	主键
title	varchar(255)		NO	名称
update_time	timestamp		NO	更新时间
value	varchar(255)		NO	数据值

2.2.10 系统用户管理表 sys_manager

系统用户管理表 sys_manager 通常用于存储系统中用户的基本信息以及与用户账户相关的配置和管理数据。这张表在大多数企业级或复杂的系统中都是不可或缺的，因为它提供了对用户身份、权限、状态等关键信息的集中管理。系统用户管理表 sys_manager 的主要功能有：

- 用户信息管理：存储用户的基本信息，如用户名、密码（通常是加密后的）、真实姓名、电子邮件、手机号码等。
- 用户状态管理：标记用户的账户状态，如启用、禁用、锁定等，以便于管理员控制用户的访问权限。
- 用户角色关联：通过外键或其他关联机制，与角色管理表（sys_role）关联，确定用户的权限范围。
- 用户登录记录：虽然可能不直接存储在 sys_manager 表中，但通常会与登录日志表关联，记录用户的登录时间、IP 地址等信息。
- 用户个性化配置：存储用户的个性化设置，如界面主题、语言偏好等。

系统用户管理表 sys_manager 的表结构如表 12 所示。

表 12 系统用户管理表的设计

字段名称	字段类型	约束	为空	字段含义
avatar	varchar(200)		NO	头像
create_time	timestamp		NO	创建时间
delete_flag	tinyint		NO	删除标识（0-正常，1-已删除）
password	varchar(100)		NO	密码
pk_id	int	主键	NO	主键
status	tinyint		NO	状态（0-停用，1-正常）
super_admin	tinyint		NO	超级管理员（0-否，1-是）
update_time	timestamp		NO	更新时间
username	varchar(50)		NO	用户名

2.2.11 角色管理表 sys_role

角色管理表 sys_role 在系统中主要用于定义和管理用户角色,这些角色通常与特定的权限集合相关联。通过角色管理,系统可以实现灵活的权限控制,使得用户可以根据其所属的角色来获得相应的系统访问和操作权限。角色管理表 sys_role 的主要功能有:

- 角色定义:存储系统中所有角色的基本信息,如角色名称、描述、状态等。
- 权限关联:通过与其他表的关联,为角色分配相应的权限。这些权限可能包括访问特定页面、执行特定操作、查看特定数据等。
- 用户关联:与用户管理表 sys_manager 关联,确定哪些用户属于某个角色,从而继承该角色的所有权限。
- 角色继承:支持角色之间的继承关系,允许高级角色继承低级角色的权限,实现权限的层级管理。

角色管理表 sys_role 的表结构如表 13 所示。

表 13 角色管理表的设计

字段名称	字段类型	约束	为空	字段含义
create_time	timestamp		NO	创建时间
delete_flag	tinyint		NO	删除标识(0-正常,1-已删除)
name	varchar(16)		NO	角色名称
pk_id	int	主键	NO	主键
remark	varchar(30)		NO	备注
update_time	timestamp		NO	更新时间

2.2.12 用户角色关系表 sys_manager_role

用户角色关系表 sys_manager_role 在系统中用于建立和维护用户与角色之间的关联关系。这张表通常用于实现基于角色的访问控制(RBAC)模型,允许系统管理员将用户分配到不同的角色中,从而为用户分配相应的权限集合。用户角色关系表 sys_manager_role 的主要功能有:

- 关联用户和角色:通过这张表,系统可以明确知道每个用户属于哪些角色,进而确定用户的权限范围。
- 支持多角色分配:一个用户可以被分配到多个角色中,从而拥有多个角色的权限集合。
- 支持灵活配置:管理员可以根据需要,随时为用户添加或删除角色,实现权限的动态管理。

用户角色关系表 sys_manager_role 的表结构如表 14 所示。

表 14 用户角色关系表的设计

字段名称	字段类型	约束	为空	字段含义
create_time	timestamp		NO	创建时间
delete_flag	tinyint		NO	删除标识(0-正常,1-已删除)
manager_id	int		NO	用户 ID
pk_id	int	主键	NO	主键
role_id	int		NO	角色 ID
update_time	timestamp		NO	更新时间

2.2.13 系统菜单表 sys_menu

系统菜单表 sys_menu 在系统中用于定义和管理系统的菜单结构。它通常包含了系统中所有菜单项的基本信息,如菜单名称、URL、父菜单 ID 等,以便于系统能够正确地展示和导航菜单。系统菜单表 sys_menu 的主要功能有:

- 菜单定义:存储系统中所有菜单项的基本信息,如菜单名称、图标、URL、排序号等。
- 菜单层级结构:通过父菜单 ID 字段,支持多级菜单的层级结构,使得系统可以灵活地组织和管理菜单项。
- 权限关联:菜单项通常与权限相关联,通过与其他表的关联,可以控制用户对菜单的访问权限。
- 动态展示:根据用户的角色和权限,动态地展示不同的菜单项,实现个性化的菜单展示。

系统菜单表 sys_menu 的表结构如表 15 所示。

表 15 系统菜单表的设计

字段名称	字段类型	约束	为空	字段含义
auth	varchar(500)		NO	授权标识(多个用逗号分隔,如:sys: menu:list,sys:menu:save)
component	varchar(200)		NO	组件路径
create_time	timestamp		NO	创建时间
delete_flag	tinyint		NO	逻辑删除(0-未删除,1-删除)
hide	tinyint		NO	是否隐藏(0-true,1-false)
icon	varchar(50)		NO	菜单图标
keep_alive	tinyint		NO	是否缓存(0-true,1-false)

（续表）

字段名称	字段类型	约束	为空	字段含义
name	varchar(200)		NO	名称
open_type	varchar(50)		NO	打开类型(tab-选项卡,url-外链)
parent_id	int		NO	父级 id
path	varchar(200)		NO	路径
pk_id	int	主键	NO	主键
sort	int		NO	排序
title	varchar(200)		NO	标题
type	varchar(50)		NO	菜单类型(menu-菜单,button-按钮)
update_time	timestamp		NO	更新时间
url	varchar(500)		NO	外链地址

2.2.14 角色菜单关系表 sys_role_menu

角色菜单关系表 sys_role_menu 在系统中用于建立和维护角色与菜单之间的关联关系。这张表的核心目的是确保每个角色都能被分配到适当的菜单权限，从而控制用户对系统菜单的访问和操作。角色菜单关系表 sys_role_menu 的主要功能有：

- 角色与菜单的关联：通过这张表，系统可以明确知道每个角色拥有哪些菜单的访问权限。
- 支持多菜单分配：一个角色可以被分配到多个菜单上，从而拥有多个菜单的访问权限。
- 动态权限管理：管理员可以根据需要，随时为角色添加或删除菜单权限，实现灵活的权限管理。

角色菜单关系表 sys_role_menu 的表结构如表 16 所示。

表 16 角色菜单关系表的设计

字段名称	字段类型	约束	为空	字段含义
create_time	timestamp		NO	创建时间
delete_flag	tinyint		NO	删除标识(0-正常,1-已删除)
menu_id	int		NO	菜单 ID
pk_id	int	主键	NO	主键
role_id	int		NO	角色 ID
update_time	timestamp		NO	更新时间

3 数据库建表脚本

3.1 任务描述

经过前期的数据表结构设计,每个表都已经有了明确的字段和约束。接下来需要将这些设计转化为实际的数据库表,这是确保系统稳定运行的关键一步。

小南作为数据库的新手,对于如何具体操作还有些许困惑。波哥耐心地解释,需要使用建表脚本语句来实现这一过程。他详细说明了建表脚本语句的作用,即它是一种特定的 SQL 语言,允许开发者根据数据表结构设计来创建数据库表。通过编写和执行这些脚本语句,可以确保数据库表能够准确地反映出设计时的意图。

建表脚本语句包括一系列的 SQL 命令,这些命令用于定义表的名称、字段、数据类型以及约束等信息。通过使用 CREATE TABLE 语句来创建新表,并使用如 VARCHAR、INT 等数据类型来定义字段,开发者可以确保数据库表能够存储正确的数据。同时,使用如 PRIMARY KEY、FOREIGN KEY 等约束条件,可以确保数据的完整性和准确性。

波哥指导着小南和小工,使用建表脚本语句不仅可以确保数据库表按照设计创建,避免错误或遗漏,还具有可重用性。这意味着一旦编写完成,这些脚本语句可以在不同的数据库环境中重复使用,极大地提高了开发效率。他鼓励小南和小工积极学习和掌握编写建表脚本语句的技巧,以便在未来的工作中能够更加熟练地操作数据库。

在波哥的指导下,小南和小工对建表脚本语句有了更深入的理解。他们意识到这是数据库开发中的一项重要技能,对于确保系统稳定运行和提高开发效率具有重要意义。因此,他们决定立即开始学习如何编写建表脚本语句,并计划在接下来的工作中积极实践这一技能。

3.2 任务分析

在数据库设计中,建表是至关重要的一环。建表脚本,作为创建数据库表的指令集合,是实现数据库表结构的重要工具。通过编写和执行建表脚本,我们能够确保数据库表按照预定的设计方案准确创建,为后续的数据存储和查询提供坚实的基础。在编写数据库建表脚本时,应该遵循以下基本要素:

- 表名:每个数据库表都需要一个唯一的名称,用于标识和引用该表。表名应遵循一定的命名规范,如使用有意义的名称、避免使用特殊字符等。
- 字段:字段是数据库表的基本组成单位,用于存储具体的数据。在建表脚本中,我们需要定义每个字段的名称、数据类型、约束等属性。字段名称应具有描述性,数据类型应根据实际需求选择,约束则用于确保数据的完整性和准确性。
- 主键:主键是数据库表中的一个特殊字段,用于唯一标识表中的每一行记录。在建表脚本中,我们需要指定主键字段,并设置相应的约束条件,以确保主键的唯一性和非空性。
- 外键:外键是数据库表之间的关联关系,用于建立表之间的引用关系。在建表脚本中,我们可以定义外键字段,并设置相应的引用关系和约束条件,以实现表之间的数据关联和引用。

因此,我们可以根据已经设计好的数据表结构,编写对应的建表脚本。

3.2.1 用户表 t_user

```
CREATE TABLE 't_user' (
  'pk_id' int NOT NULL AUTO_INCREMENT COMMENT '主键',
  'phone' varchar(11) CHARACTER SET utf8mb4 COLLATE utf8mb4_general_ci DEFAULT NULL COMMENT '手机号',
  'wx_open_id' varchar(255) CHARACTER SET utf8mb4 COLLATE utf8mb4_general_ci DEFAULT NULL COMMENT '微信 openId',
  'nickname' varchar(36) CHARACTER SET utf8mb4 COLLATE utf8mb4_general_ci NOT NULL COMMENT '昵称',
  'avatar' varchar(1000) CHARACTER SET utf8mb4 COLLATE utf8mb4_general_ci NOT NULL COMMENT '头像',
  'gender' tinyint DEFAULT NULL COMMENT '性别(0-男,1-女)',
  'birthday' varchar(255) CHARACTER SET utf8mb4 COLLATE utf8mb4_general_ci DEFAULT NULL COMMENT '生日',
  'bonus' int DEFAULT NULL COMMENT '积分',
  'remark' varchar(255) CHARACTER SET utf8mb4 COLLATE utf8mb4_general_ci DEFAULT NULL COMMENT '备用字段',
  'enabled' tinyint NOT NULL COMMENT '是否可用(0-冻结,1-可用)',
  'delete_flag' tinyint NOT NULL COMMENT '删除标识(0-未删除,1-删除)',
  'update_time' timestamp NOT NULL DEFAULT CURRENT_TIMESTAMP ON UPDATE CURRENT_TIMESTAMP COMMENT '更新时间',
  'create_time' timestamp NOT NULL DEFAULT CURRENT_TIMESTAMP COMMENT '创建时间',
  PRIMARY KEY ('pk_id') USING BTREE
) ENGINE=InnoDB AUTO_INCREMENT=11 DEFAULT CHARSET=utf8mb4 COLLATE=utf8mb4_general_ci ROW_FORMAT=DYNAMIC;
```

3.2.2 用户行为表 t_action

```
CREATE TABLE 't_action' (
  'pk_id' int NOT NULL AUTO_INCREMENT COMMENT '主键',
```

'user_id' int NOT NULL COMMENT '用户id',

'resource_id' int NOT NULL COMMENT '资源id',

'type' tinyint NOT NULL COMMENT '类型(0-收藏,1-发布,2-兑换,3-点赞)',

'delete_flag' tinyint NOT NULL COMMENT '删除标识(0-未删除,1-删除)',

'update_time' timestamp NOT NULL DEFAULT CURRENT_TIMESTAMP ON UPDATE CURRENT_TIMESTAMP COMMENT '更新时间',

'create_time' timestamp NOT NULL DEFAULT CURRENT_TIMESTAMP COMMENT '创建时间',

PRIMARY KEY ('pk_id') USING BTREE

) ENGINE=InnoDB AUTO_INCREMENT=25 DEFAULT CHARSET=utf8mb4 COLLATE=utf8mb4_general_ci ROW_FORMAT=DYNAMIC;

3.2.3 积分日志表 t_bonus_log

CREATE TABLE 't_bonus_log' (

'pk_id' int NOT NULL AUTO_INCREMENT COMMENT '主键',

'user_id' int NOT NULL COMMENT '用户id',

'content' varchar(255) CHARACTER SET utf8mb4 COLLATE utf8mb4_general_ci NOT NULL COMMENT '用户行为',

'bonus' int NOT NULL COMMENT '积分',

'delete_flag' tinyint NOT NULL COMMENT '删除标识(0-未删除,1-删除)',

'update_time' timestamp NOT NULL DEFAULT CURRENT_TIMESTAMP ON UPDATE CURRENT_TIMESTAMP COMMENT '更新时间',

'create_time' timestamp NOT NULL DEFAULT CURRENT_TIMESTAMP COMMENT '创建时间',

PRIMARY KEY ('pk_id') USING BTREE

) ENGINE=InnoDB AUTO_INCREMENT=21 DEFAULT CHARSET=utf8mb4 COLLATE=utf8mb4_general_ci ROW_FORMAT=DYNAMIC;

3.2.4 资源分类表 t_category

CREATE TABLE 't_category' (

'pk_id' int NOT NULL AUTO_INCREMENT COMMENT '主键',

'title' varchar(255) CHARACTER SET utf8mb4 COLLATE utf8mb4_general_ci NOT NULL COMMENT '分类名称',

'type' tinyint NOT NULL COMMENT '分类类型(0-网盘类型,1-资源类型)',

'description' varchar(255) CHARACTER SET utf8mb4 COLLATE utf8mb4_general_ci DEFAULT NULL COMMENT '描述',

'delete_flag' tinyint NOT NULL COMMENT '删除标识(0-未删除,1-删除)',

`update_time` timestamp NOT NULL DEFAULT CURRENT_TIMESTAMP ON UPDATE CURRENT_TIMESTAMP COMMENT '更新时间',
　　`create_time` timestamp NOT NULL DEFAULT CURRENT_TIMESTAMP COMMENT '创建时间',
　　PRIMARY KEY (`pk_id`) USING BTREE
) ENGINE=InnoDB AUTO_INCREMENT=17 DEFAULT CHARSET=utf8mb4 COLLATE=utf8mb4_general_ci ROW_FORMAT=DYNAMIC;

3.2.5　资源表 t_resource

CREATE TABLE `t_resource` (
　　`pk_id` int NOT NULL AUTO_INCREMENT COMMENT '主键',
　　`title` varchar(255) CHARACTER SET utf8mb4 COLLATE utf8mb4_general_ci NOT NULL COMMENT '标题',
　　`author` int NOT NULL COMMENT '发布人 id',
　　`disk_type` int NOT NULL COMMENT '网盘分类 id',
　　`res_type` json NOT NULL COMMENT '资源分类 id,多个',
　　`tags` json NOT NULL COMMENT '标签',
　　`download_url` varchar(255) CHARACTER SET utf8mb4 COLLATE utf8mb4_general_ci NOT NULL COMMENT '资源链接',
　　`detail` text CHARACTER SET utf8mb4 COLLATE utf8mb4_general_ci NOT NULL COMMENT '详情',
　　`price` int NOT NULL COMMENT '价格',
　　`like_num` int NOT NULL COMMENT '点赞量',
　　`is_top` tinyint NOT NULL COMMENT '是否置顶(0-否,1-是)',
　　`status` tinyint NOT NULL COMMENT '审核状态(0-待审核,1-通过,2-拒绝)',
　　`remark` varchar(255) CHARACTER SET utf8mb4 COLLATE utf8mb4_general_ci DEFAULT NULL COMMENT '审核结果描述',
　　`delete_flag` tinyint NOT NULL COMMENT '删除标识(0-未删除,1-删除)',
　　`update_time` timestamp NOT NULL DEFAULT CURRENT_TIMESTAMP ON UPDATE CURRENT_TIMESTAMP COMMENT '更新时间',
　　`create_time` timestamp NOT NULL DEFAULT CURRENT_TIMESTAMP COMMENT '创建时间',
　　PRIMARY KEY (`pk_id`) USING BTREE
) ENGINE=InnoDB AUTO_INCREMENT=19 DEFAULT CHARSET=utf8mb4 COLLATE=utf8mb4_general_ci ROW_FORMAT=DYNAMIC;

3.2.6　标签表 t_tag

CREATE TABLE `t_tag` (
　　`pk_id` int NOT NULL AUTO_INCREMENT COMMENT '主键',

'title' varchar(255) CHARACTER SET utf8mb4 COLLATE utf8mb4_general_ci DEFAULT NULL COMMENT '标签名',

'description' varchar(255) CHARACTER SET utf8mb4 COLLATE utf8mb4_general_ci DEFAULT NULL COMMENT '描述',

'is_hot' tinyint NOT NULL COMMENT '是否热门(0-否,1-是)',

'delete_flag' tinyint NOT NULL COMMENT '删除标识(0-未删除,1-删除)',

'update_time' timestamp NOT NULL DEFAULT CURRENT_TIMESTAMP ON UPDATE CURRENT_TIMESTAMP COMMENT '更新时间',

'create_time' timestamp NOT NULL DEFAULT CURRENT_TIMESTAMP COMMENT '创建时间',

PRIMARY KEY ('pk_id') USING BTREE

) ENGINE=InnoDB AUTO_INCREMENT=27 DEFAULT CHARSET=utf8mb4 COLLATE=utf8mb4_general_ci ROW_FORMAT=DYNAMIC;

3.2.7 公告表 t_notice

CREATE TABLE 't_notice' (

'pk_id' int NOT NULL AUTO_INCREMENT COMMENT '主键',

'title' varchar(255) CHARACTER SET utf8mb4 COLLATE utf8mb4_general_ci NOT NULL COMMENT '标题',

'cover' varchar(255) CHARACTER SET utf8mb4 COLLATE utf8mb4_general_ci DEFAULT NULL COMMENT '封面图',

'content' varchar(255) CHARACTER SET utf8mb4 COLLATE utf8mb4_general_ci NOT NULL COMMENT '内容',

'is_top' tinyint NOT NULL COMMENT '是否置顶(0-否,1-是)',

'is_swiper' tinyint NOT NULL COMMENT '是否轮播(0-否,1-是)',

'delete_flag' tinyint NOT NULL COMMENT '删除标识(0-未删除,1-删除)',

'update_time' timestamp NOT NULL DEFAULT CURRENT_TIMESTAMP ON UPDATE CURRENT_TIMESTAMP COMMENT '更新时间',

'create_time' timestamp NOT NULL DEFAULT CURRENT_TIMESTAMP COMMENT '创建时间',

PRIMARY KEY ('pk_id') USING BTREE

) ENGINE=InnoDB AUTO_INCREMENT=8 DEFAULT CHARSET=utf8mb4 COLLATE=utf8mb4_general_ci ROW_FORMAT=DYNAMIC;

3.2.8 系统字典表 sys_dict

CREATE TABLE 'sys_dict' (

'pk_id' int NOT NULL AUTO_INCREMENT COMMENT '主键',

'title' varchar(255) CHARACTER SET utf8mb3 COLLATE utf8mb3_general_ci NOT NULL DEFAULT '' COMMENT '名称',

'number' varchar(64) CHARACTER SET utf8mb3 COLLATE utf8mb3_general_ci NOT NULL DEFAULT '' COMMENT '编号',

'description' varchar(255) CHARACTER SET utf8mb3 COLLATE utf8mb3_general_ci NOT NULL DEFAULT '' COMMENT '描述',

'delete_flag' tinyint NOT NULL DEFAULT '0' COMMENT '删除标识(0-未删除,1-已删除)',

'create_time' timestamp NOT NULL DEFAULT CURRENT_TIMESTAMP COMMENT '创建时间',

'update_time' timestamp NOT NULL DEFAULT CURRENT_TIMESTAMP ON UPDATE CURRENT_TIMESTAMP COMMENT '更新时间',

PRIMARY KEY ('pk_id') USING BTREE

) ENGINE=InnoDB AUTO_INCREMENT=7 DEFAULT CHARSET=utf8mb3 ROW_FORMAT=DYNAMIC;

3.2.9 系统字典配置表 sys_dict_config

CREATE TABLE 'sys_dict_config' (

'pk_id' int NOT NULL AUTO_INCREMENT COMMENT '主键',

'dict_number' varchar(64) CHARACTER SET utf8mb3 COLLATE utf8mb3_general_ci NOT NULL COMMENT '字典编号',

'title' varchar(255) CHARACTER SET utf8mb3 COLLATE utf8mb3_general_ci NOT NULL COMMENT '名称',

'value' varchar(255) CHARACTER SET utf8mb3 COLLATE utf8mb3_general_ci NOT NULL COMMENT '数据值',

'delete_flag' tinyint NOT NULL DEFAULT '0' COMMENT '删除标识(0-未删除,1-已删除)',

'create_time' timestamp NOT NULL DEFAULT CURRENT_TIMESTAMP COMMENT '创建时间',

'update_time' timestamp NOT NULL DEFAULT CURRENT_TIMESTAMP ON UPDATE CURRENT_TIMESTAMP COMMENT '更新时间',

PRIMARY KEY ('pk_id') USING BTREE

) ENGINE=InnoDB AUTO_INCREMENT=14 DEFAULT CHARSET=utf8mb3 ROW_FORMAT=DYNAMIC;

3.2.10 系统用户管理表 sys_manager

CREATE TABLE 'sys_manager' (

'pk_id' int NOT NULL AUTO_INCREMENT COMMENT '主键',

'username' varchar(50) CHARACTER SET utf8mb4 COLLATE utf8mb4_general_ci NOT NULL DEFAULT '' COMMENT '用户名',

'password' varchar(100) CHARACTER SET utf8mb4 COLLATE utf8mb4_general_ci NOT NULL DEFAULT '' COMMENT '密码',

'avatar' varchar(200) CHARACTER SET utf8mb4 COLLATE utf8mb4_general_ci NOT NULL DEFAULT '' COMMENT '头像',

'super_admin' tinyint NOT NULL DEFAULT '0' COMMENT '超级管理员(0-否,1-是)',

'status' tinyint NOT NULL DEFAULT '1' COMMENT '状态(0-停用,1-正常)',

'delete_flag' tinyint NOT NULL DEFAULT '0' COMMENT '删除标识(0-正常,1-已删除)',

'create_time' timestamp NOT NULL DEFAULT CURRENT_TIMESTAMP COMMENT '创建时间',

'update_time' timestamp NOT NULL DEFAULT CURRENT_TIMESTAMP ON UPDATE CURRENT_TIMESTAMP COMMENT '更新时间',

PRIMARY KEY ('pk_id') USING BTREE

) ENGINE=InnoDB AUTO_INCREMENT=3 DEFAULT CHARSET=utf8mb4 COLLATE=utf8mb4_general_ci ROW_FORMAT=DYNAMIC COMMENT='用户管理';

3.2.11 角色管理表 sys_role

CREATE TABLE 'sys_role' (

'pk_id' int NOT NULL AUTO_INCREMENT COMMENT '主键',

'name' varchar(16) CHARACTER SET utf8mb4 COLLATE utf8mb4_general_ci NOT NULL DEFAULT '' COMMENT '角色名称',

'remark' varchar(30) CHARACTER SET utf8mb4 COLLATE utf8mb4_general_ci NOT NULL DEFAULT '' COMMENT '备注',

'delete_flag' tinyint NOT NULL DEFAULT '0' COMMENT '删除标识(0-正常,1-已删除)',

'create_time' timestamp NOT NULL DEFAULT CURRENT_TIMESTAMP COMMENT '创建时间',

'update_time' timestamp NOT NULL DEFAULT CURRENT_TIMESTAMP COMMENT '更新时间',

PRIMARY KEY ('pk_id') USING BTREE

) ENGINE=InnoDB AUTO_INCREMENT=3 DEFAULT CHARSET=utf8mb4 COLLATE=utf8mb4_general_ci ROW_FORMAT=DYNAMIC COMMENT='角色管理';

3.2.12 用户角色关系表 sys_manager_role

CREATE TABLE 'sys_manager_role' (

'pk_id' int NOT NULL AUTO_INCREMENT COMMENT '主键',

'role_id' int NOT NULL COMMENT '角色ID',

'manager_id' int NOT NULL COMMENT '用户ID',

'delete_flag' tinyint NOT NULL DEFAULT '0' COMMENT '删除标识(0-正常,1-已删除)',

'create_time' timestamp NOT NULL DEFAULT CURRENT_TIMESTAMP COMMENT '创建时间',

'update_time' timestamp NOT NULL DEFAULT CURRENT_TIMESTAMP ON UPDATE CURRENT_TIMESTAMP COMMENT '更新时间',

```
  PRIMARY KEY ('pk_id') USING BTREE
) ENGINE=InnoDB AUTO_INCREMENT=3 DEFAULT CHARSET=utf8mb4 COLLATE=utf8mb4_general_ci ROW_FORMAT=DYNAMIC COMMENT='用户角色关系';
```

3.2.13 系统菜单表 sys_menu

```
CREATE TABLE 'sys_menu' (
  'pk_id' int NOT NULL AUTO_INCREMENT COMMENT '主键',
  'parent_id' int NOT NULL DEFAULT '0' COMMENT '父级 id',
  'name' varchar(200) CHARACTER SET utf8mb3 COLLATE utf8mb3_general_ci NOT NULL DEFAULT '' COMMENT '名称',
  'title' varchar(200) CHARACTER SET utf8mb3 COLLATE utf8mb3_general_ci NOT NULL DEFAULT '' COMMENT '标题',
  'path' varchar(200) CHARACTER SET utf8mb3 COLLATE utf8mb3_general_ci NOT NULL DEFAULT '' COMMENT '路径',
  'component' varchar(200) CHARACTER SET utf8mb3 COLLATE utf8mb3_general_ci NOT NULL DEFAULT '' COMMENT '组件路径',
  'type' varchar(50) CHARACTER SET utf8mb3 COLLATE utf8mb3_general_ci NOT NULL DEFAULT '' COMMENT '菜单类型(menu-菜单,button-按钮)',
  'open_type' varchar(50) CHARACTER SET utf8mb3 COLLATE utf8mb3_general_ci NOT NULL DEFAULT '' COMMENT '打开类型(tab-选项卡,url-外链)',
  'url' varchar(500) CHARACTER SET utf8mb3 COLLATE utf8mb3_general_ci NOT NULL DEFAULT '' COMMENT '外链地址',
  'icon' varchar(50) CHARACTER SET utf8mb3 COLLATE utf8mb3_general_ci NOT NULL DEFAULT '' COMMENT '菜单图标',
  'auth' varchar(500) CHARACTER SET utf8mb3 COLLATE utf8mb3_general_ci NOT NULL DEFAULT '' COMMENT '授权标识(多个用逗号分隔,如:sys:menu:list,sys:menu:save)',
  'keep_alive' tinyint NOT NULL DEFAULT '0' COMMENT '是否缓存(0-true,1-false)',
  'sort' int NOT NULL DEFAULT '0' COMMENT '排序',
  'delete_flag' tinyint NOT NULL DEFAULT '0' COMMENT '逻辑删除(0-未删除,1-删除)',
  'create_time' timestamp NOT NULL DEFAULT CURRENT_TIMESTAMP COMMENT '创建时间',
  'update_time' timestamp NOT NULL DEFAULT CURRENT_TIMESTAMP ON UPDATE CURRENT_TIMESTAMP COMMENT '更新时间',
  'hide' tinyint NOT NULL DEFAULT '1' COMMENT '是否隐藏(0-true,1-fasle)',
  PRIMARY KEY ('pk_id') USING BTREE
) ENGINE=InnoDB AUTO_INCREMENT=144 DEFAULT CHARSET=utf8mb3 ROW_FORMAT=DYNAMIC;
```

3.2.14 角色菜单关系表 sys_role_menu

```
CREATE TABLE 'sys_role_menu' (
  'pk_id' int NOT NULL AUTO_INCREMENT COMMENT '主键',
  'role_id' int NOT NULL COMMENT '角色 ID',
  'menu_id' int NOT NULL COMMENT '菜单 ID',
  'delete_flag' tinyint NOT NULL DEFAULT '0' COMMENT '删除标识(0-正常,1-已删除)',
  'create_time' timestamp NOT NULL DEFAULT CURRENT_TIMESTAMP COMMENT '创建时间',
  'update_time' timestamp NOT NULL DEFAULT CURRENT_TIMESTAMP ON UPDATE CURRENT_TIMESTAMP COMMENT '更新时间',
  PRIMARY KEY ('pk_id') USING BTREE
) ENGINE=InnoDB AUTO_INCREMENT=86 DEFAULT CHARSET=utf8mb4 COLLATE=utf8mb4_general_ci ROW_FORMAT=DYNAMIC COMMENT='角色菜单关系';
```

4 任务总结

本次任务主要完成了数据库设计、数据表结构设计和数据库建表脚本三个关键阶段。

1. 数据库设计：

数据库设计是数据库系统开发的起始阶段，它涉及确定数据库系统的目标、范围、功能需求和非功能需求。

设计过程通常包括概念设计（ER 图建模）、逻辑设计（将概念设计转化为数据库管理系统可识别的逻辑结构）和物理设计（确定数据的存储结构和访问方法）。

2. 数据表结构设计：

数据表结构设计是数据库逻辑设计的一部分，它专注于定义数据库中的各个表以及这些表之间的关系。

在这个阶段，我们需要确定每个表的字段、字段的数据类型、约束（如主键、外键、唯一性约束等）、索引等。

一个良好的数据表结构设计应该满足数据的完整性、一致性、安全性和性能要求。

3. 数据库建表脚本：

数据库建表脚本是根据数据表结构设计编写的 SQL 语句集合，用于在数据库管理系统中实际创建数据表。

建表脚本通常包括 CREATE TABLE 语句，用于定义表名、字段名、数据类型、约束等。

通过执行建表脚本，我们可以自动化地创建数据库表，并确保表的结构与我们的设计保持一致。

本章介绍了从概念到实际实现的数据库开发过程，强调了数据库设计、数据表结构设计和数据库建表脚本的重要性,并提供了相应的指导和建议。我们通过合理的数据库设计和数据表结构设计,以及准确的建表脚本编写,为数据库系统的稳定运行和高效查询奠定坚实的基础。

扫描二维码,参考文件夹中"任务一完整SQL"文件。

任务二

设计接口文档

波哥带领项目团队已经顺利完成了数据库的设计工作。现在,他们准备进入接口文档设计的阶段。

波哥召集了团队成员:"各位,接下来我们的任务是设计接口文档。这不仅是前后端开发人员沟通的关键,也是确保项目顺利推进的重要环节。"

◇ 任务点

- 编写客户端 API 接口文档;
- 编写后台管理系统 API 接口文档;
- 掌握接口文档的编写规范和技巧。

◇ 任务计划

- 任务内容:设计接口并编写接口文档;
- 任务耗时:预计完成时间为 2~3 h;
- 任务难点:合理地设计 API 接口。

1 后台管理系统接口文档

1.1 任务描述

波哥召集了小南和小工进行了一次关于接口文档设计的讨论。

波哥首先强调:"为了确保后台管理系统与前端及其他系统模块之间的顺畅通信,我们需要设计一份详尽的接口文档。"他随后详细阐述了接口文档的重要性,并指出这是确保系统间通信一致性和准确性的关键。

小南作为团队成员之一,对于如何设计接口文档表达了疑虑,不知道应该如何开始设计接口文档。

波哥耐心地解释着:要明确后台管理系统需要提供哪些功能接口,如用户管理、数据查询、报表生成等。然后,为每个接口定义清晰的请求和响应格式,包括请求方法、参数、状态码和响应数据等。

在讨论过程中,小工也积极参与,并提出了关于接口设计规范的疑问:"我们在设计接口时,有哪些规范需要遵循吗?"

波哥回答道:"当然,我们需要确保接口设计的一致性,遵循 RESTful 架构原则,并使用合适的 HTTP 方法。同时,安全性也是不可忽视的,建议使用 HTTPS 协议进行加密传输,并实施身份验证和授权机制。在接口文档中,要注明每个接口的限制和注意事项,如请求频率限制、参数的必填项和可选项等,以确保前端开发人员能够准确理解并遵循。"

通过波哥的指导和小南、小工的积极参与,团队明确了接口文档的设计方向和要求。他们意识到,通过对话和客观描述的集合方式,可以更好地理解和把握接口设计的关键要点,为后续的开发工作打下坚实的基础。

1.2 任务分析

接口文档是系统开发中不可或缺的一部分,它详细描述了系统提供的各种接口的功能、使用方法、请求参数、响应格式等信息。对于后台管理系统而言,接口文档的重要性主要体现在以下几个方面:

- 提高开发效率:通过接口文档,开发人员可以清晰地了解后台管理系统提供了哪些接口,以及如何使用这些接口。这有助于减少开发过程中的沟通和误解,提高开发效率。
- 保证数据一致性:接口文档规定了数据的请求和响应格式,以及数据的处理和验证规则。这有助于确保不同系统模块之间数据的一致性,避免数据格式错误或数据丢失等问题。
- 支持团队协作:接口文档是团队协作的基础。不同开发团队(如前端、后端、测试等)可以根据接口文档进行并行开发,减少等待和依赖,提高团队协作效率。

一个完整的接口文档通常包含以下内容:

- 接口概述:介绍接口的目的、功能、使用场景等基本信息。
- 接口请求:详细描述接口的请求方法(如 GET、POST 等)、请求 URL、请求参数(包括必填项和可选项)、请求头等信息。
- 接口响应:描述接口的响应状态码、响应数据格式、响应示例等信息。对于可能出现的错误或异常情况,也需要进行说明。
- 接口安全性:介绍接口的安全性措施,如身份验证、授权机制、加密传输等。
- 接口限制与注意事项:注明接口的请求频率限制、参数限制等限制条件,以及使用接口时需要注意的事项。

1.2.1 接口设计说明

本系统所有 API 接口采用 RESTful 风格设计,主要 HTTP 方法包含:GET、POST。接口访问举例如下:

1. GET 方法

GET 请求拼接 URL 参数：/share-admin-api/user/bonus/list

请求参数如表1所示。

表1 请求参数（GET 方法）

参数名称	参数说明	请求类型	是否必须	数据类型	schema
userId		query	true	integer(int32)	

请求示例：GET/share-admin-api/user/bonus/list? userId=1

请求解释：获取用户 ID 为 1 的用户积分数据。

2. POST 方法

POST 请求拼接 URL 路径：POST/share-admin-api/sys/auth/login

请求参数如表2所示。

表2 请求参数（POST 方法）

参数名称	参数说明	请求类型	是否必须	数据类型	schema
sysAccountLoginVO	账号登录	body	true	SysAccountLoginVO	SysAccountLoginVO
username	用户名		false	string	
password	密码		false	string	

请求示例

{
"username"："admin"，
"password"："admin123"
}

请求解释：以用户名"admin"和密码"admin123"进行登录

1.2.2 接口列表

1.2.2.1 管理员管理接口（表3）

表3 管理员管理接口

接口描述	接口地址	请求方式
新增	/share-admin-api/sys/manager/add	POST
修改密码	/share-admin-api/sys/manager/changePassword	POST
修改	/share-admin-api/sys/manager/edit	POST
获取管理员信息	/share-admin-api/sys/manager/getManagerInfo	POST
分页	/share-admin-api/sys/manager/page	POST
删除	/share-admin-api/sys/manager/remove	POST

1.2.2.2 角色管理接口（表4）

表4　角色管理接口

接口描述	接口地址	请求方式
新增	/share-admin-api/sys/role/add	POST
修改	/share-admin-api/sys/role/edit	POST
列表	/share-admin-api/sys/role/list	POST
角色表单菜单列表	/share-admin-api/sys/role/menu	POST
分页	/share-admin-api/sys/role/page	POST
删除	/share-admin-api/sys/role/remove	POST

1.2.2.3 菜单管理接口（表5）

表5　菜单管理接口

接口描述	接口地址	请求方式
新增	/share-admin-api/sys/menu/add	POST
用户按钮权限	/share-admin-api/sys/menu/button	POST
修改	/share-admin-api/sys/menu/edit	POST
表单菜单列表	/share-admin-api/sys/menu/form	POST
菜单信息	/share-admin-api/sys/menu/info	POST
菜单列表	/share-admin-api/sys/menu/list	POST
用户菜单	/share-admin-api/sys/menu/nav	POST
删除	/share-admin-api/sys/menu/remove	POST

1.2.2.4 字典管理接口（表6）

表6　字典管理接口

接口描述	接口地址	请求方式
修改字典	/share-admin-api/dict/edit	POST
字典列表	/share-admin-api/dict/page	POST
删除字典	/share-admin-api/dict/remove	POST
新增字典	/share-admin-api/dict/save	POST

1.2.2.5 字典配置管理接口(表7)

表7 字典配置管理接口

接口描述	接口地址	请求方式
修改字典配置	/share-admin-api/dictConfig/edit	POST
字典配置列表	/share-admin-api/dictConfig/page	POST
删除字典配置	/share-admin-api/dictConfig/remove	POST
新增字典配置	/share-admin-api/dictConfig/save	POST

1.2.2.6 认证管理接口(表8)

表8 认证管理接口

接口描述	接口地址	请求方式
账号密码登录	/share-admin-api/sys/auth/login	POST
退出登录	/share-admin-api/sys/auth/logout	POST

1.2.2.7 分类管理接口(表9)

表9 分类管理接口

接口描述	接口地址	请求方式
分页查询分类	/share-admin-api/category/page	POST
新增或修改	/share-admin-api/category/saveAndEdit	POST
删除	/share-admin-api/category/delete	POST

1.2.2.8 首页接口(表10)

表10 首页接口

接口描述	接口地址	请求方式
欢迎	/share-admin-api/index	GET
首页数据	/share-admin-api/index/dashboard	GET

1.2.2.9 通用模块接口(表11)

表11 通用模块接口

接口描述	接口地址	请求方式
图片上传	/share-admin-api/common/upload/img	POST

1.2.2.10 用户管理接口(表12)

表12 用户管理接口

接口描述	接口地址	请求方式
积分列表	/share-admin-api/user/bonus/list	GET
修改	/share-admin-api/user/edit	POST
账户状态修改	/share-admin-api/user/enabled	POST
导出	/share-admin-api/user/export	POST
全部用户列表	/share-admin-api/user/list	GET
分页	/share-admin-api/user/page	GET

1.2.2.11 资源管理接口(表13)

表13 资源管理接口

接口描述	接口地址	请求方式
分页查询资源	/share-admin-api/resource/page	POST
审核资源	/share-admin-api/resource/audit	POST

1.2.2.12 标签管理接口(表14)

表14 标签管理接口

接口描述	接口地址	请求方式
分页查询公告	/share-admin-api/tag/page	POST
新增或修改	/share-admin-api/tag/saveAndEdit	POST
删除	/share-admin-api/tag/delete	POST

1.2.2.13 公告管理接口(表15)

表15 公告管理接口

接口描述	接口地址	请求方式
分页查询公告	/share-admin-api/notice/page	POST
新增或修改	/share-admin-api/notice/saveAndEdit	POST
删除	/share-admin-api/notice/delete	POST

2 任务总结

本次任务主要是根据功能需求设计了后台管理系统API接口,并编写了接口文档。

接口文档是确保后台管理系统与其他系统模块顺畅通信的关键,它详细描述了后台管理系统提供的各种接口的功能、使用方法、请求参数、响应格式等信息。通过接口文档,开发人员可以清晰地了解后台管理系统提供的接口,并据此进行开发。这有助于提高开发效率,减少沟通误解,保证数据的一致性,并支持团队协作。接口文档通常包含接口概述、请求、响应、安全性、限制与注意事项等内容。在编写接口文档时,需要注意清晰明了、准确完整、易于阅读以及及时更新等要点,以确保文档的质量和有效性。

　　扫描二维码,参考文件夹中"任务二客户端小程度接口文档"和"任务二管理后台接口文档"两个文件。

任务三

搭建后台管理系统

波哥带领的项目团队在完成了数据库和接口设计后,开始转向了后台管理系统的搭建工作。

"各位,我们现在要进入后台管理系统的搭建阶段了。"波哥郑重其事地说道,"后台管理系统是我们项目的核心组成部分,它将直接负责数据的管理和维护,确保系统的稳定运行。"

"任务三到任务十是项目中的关键部分,涉及前后端的紧密协作。大家需要确保各自的工作能够顺利对接,实现功能的完整性和稳定性,我相信你们一定能够顺利完成这些任务。"

◇ 任务点

- 后台管理系统后端搭建;
- 后台管理系统前端搭建。

◇ 任务计划

- 任务内容:后台管理系统的搭建以及简单接口测试运行;
- 任务耗时:预计完成时间为 1~2 h;
- 任务难点:熟练掌握后台管理系统的操作使用。

1 后台管理系统后端

1.1 任务描述

波哥站在会议室前,目光扫过小南和小工:"各位,现在我们即将一起开始搭建后台管理系统的后端。在这个过程中,有几个关键步骤需要我们的共同关注。

首先,我们要搭建一个高效的开发环境,选择合适的开发工具对于提高开发效率至关重要。这些工具不仅能快速构建、调试和测试代码,还能提供代码自动补全、版本控制等强大功能,帮助我们更加高效地完成工作。

接下来,我们将从下载项目模版开始,为项目搭建一个坚实的基础。项目模版包含了预先定义好的项目结构、配置文件和部分代码,能够为我们节省大量的时间和精力。通过选择适合的模版,我们可以迅速构建起项目的基本框架和所需的功能模块。

在了解项目模版后,我们需要对项目的功能模块有一个清晰的认识。每个模块都承载着特定的功能和职责,它们通过定义好的接口相互协作,共同实现整个系统的功能。因此,熟悉项目结构和模块之间的关联关系,对于后续的开发工作至关重要。

在启动项目之前,我们还需要确保开发环境已经配置好。这包括选择适合的操作系统、Web 服务器、Java 开发包和数据库等。同时,我们还需要选择一款合适的开发工具,如 IntelliJ IDEA 或 Visual Studio Code,以提高我们的开发效率。

最后,在项目开发过程中和完成后,我们需要对项目进行功能测试。功能测试是确保系统稳定和正确的重要步骤,通过模拟用户操作,对系统的各项功能进行详细的测试。这将帮助我们及时发现潜在的问题并进行修复,确保系统在实际使用中能够正常运行。"

1.2 任务分析

1.2.1 开发环境搭建

在进行后端开发的过程中,选择一款合适的开发工具能够极大地提高开发效率。开发工具不仅能帮助我们快速构建、调试和测试代码,还能提供一系列强大的功能,如代码自动补全、版本控制、代码审查等。接下来,我们将介绍几款常用的后端开发工具,帮助读者更好地进行后端开发。

1. 集成开发环境(IDE)

集成开发环境(IDE)是后端开发中最常用的工具之一。它提供了一个集成的开发平台,包含了代码编辑、编译、调试、版本控制等功能。以下是几款常用的 IDE:

(1) Visual Studio Code:由微软开发的轻量级、跨平台的 IDE。它支持多种编程语言,包括 JavaScript、Python、Java 等,并提供了丰富的插件生态,可以方便地扩展功能。

(2) IntelliJ IDEA:由 JetBrains 公司开发的强大的 Java IDE。它支持多种 Java 框架和库,并提供了智能代码补全、代码重构、代码审查等高级功能。

(3) Eclipse:Eclipse 是一款基于 Java 的开源、可扩展的集成开发环境(IDE)。它最初由 IBM 开发,后由 Eclipse 基金会管理。Eclipse 支持多种编程语言,如 Java、C/C++、PHP、Python 等,并通过插件系统支持其他语言和工具。

2. 版本控制系统

版本控制系统是后端开发中不可或缺的工具。它可以帮助我们追踪代码的变更历史,管理不同版本的代码,并协同团队成员进行代码开发。以下是几款常用的版本控制系统:

(1) Git:目前最流行的版本控制系统之一。它支持分布式版本控制,具有灵活、高效、可扩展等特点。GitHub 和 GitLab 等平台提供了 Git 代码托管和协作开发的服务。

(2) SVN(Subversion):较早的版本控制系统之一。它采用集中式版本控制模型,适合小团队或项目使用。

3. 代码编辑器

除了 IDE 外,还有一些轻量级的代码编辑器也备受开发者喜爱。它们通常具有快速启动、占用资源少等特点,适合进行简单的代码编辑和查看。以下是一些常用的代码编辑器:

(1) Sublime Text:一款高度可定制的文本编辑器,支持多种编程语言,并提供了丰富的插件生态。

(2) Atom:由 GitHub 开发的开源文本编辑器。它基于 Web 技术构建,具有现代化的界面和丰富的插件支持。

4. 构建和部署工具

在开发过程中,我们需要构建和部署代码到生产环境。以下是一些常用的构建和部署工具:

(1) Maven:Java 项目的构建和依赖管理工具。它使用 XML 文件来描述项目信息,并可以自动下载和管理项目所需的依赖库。

(2) Gradle:另一款 Java 项目的构建工具。它使用 Groovy 语言编写构建脚本,具有更灵活和强大的功能。

(3) Docker:容器化技术的代表。它可以将应用程序及其运行环境打包成一个独立的容器,方便在不同的环境中进行部署和运行。

1.2.2 下载项目模板

在系统开发过程中,为了快速搭建起项目的基本框架和所需的功能模块,通常会使用项目模板作为起点。项目模板包含了预先定义好的项目结构、配置文件以及部分代码,可以帮助开发者节省大量的时间和精力。

项目模板仓库通常保存于在线代码托管平台,如 GitHub、GitLab 或 Bitbucket 等。如果使用 Git 作为版本控制系统,可以直接在命令行中使用 git clone 命令克隆模板仓库。如果不使用 Git,也可以通过点击仓库中的"Download"按钮或链接,将模板下载到本地。如果下载的模板是压缩包(如.zip、.tar.gz 等),还需要将其解压到本地的一个目录中。解压后,你将看到整个项目的目录结构和文件。

扫描二维码,参考文件夹中"任务三后台管理系统后端项目模板"。

1.2.3 项目功能模块介绍

系统的项目结构如图 1 所示。

```
∨ 📁 src
  ∨ 📁 main
    ∨ 📁 java
      ∨ 📁 top.ssy.share.admin
        > 📁 common
        > 📁 controller
        > 📁 convert
        > 📁 converter
        > 📁 enums
        > 📁 mapper
        ∨ 📁 model
          > 📁 dto
          > 📁 entity
          > 📁 query
          > 📁 vo
        ∨ 📁 security
          > 📁 cache
          > 📁 config
          > 📁 exception
          > 📁 filter
          > 📁 service
          > 📁 user
          > 📁 utils
        > 📁 serializer
        > 📁 service
        > 📁 utils
        🅖 ShareAdminApiApplication
  ∨ 📁 resources
    > 📁 log
    > 📁 mapper
    🔧 application.yml
    🔧 application-dev.yml
```

图 1　系统项目结构

java 目录下是系统代码，resources 目录下是系统配置文件，每个目录具体含义如表 1 所示。

表 1 系统目录文件内容

目录名称	子目录	含义
common		公共目录，存放项目中通用的工具类、配置类、常量以及异常处理等内容
	cache	缓存相关代码，本系统中主要与 redis 相关
	config	存放项目中常用的配置文件，如：MyBatis 分页配置等
	constant	基础的常量声明
	exception	声明自定义异常，以及全局异常处理
	handler	分离项目中通用的业务逻辑
	result	封装接口统一响应体
	interceptor	存放拦截器相关的配置，拦截器可以在请求被处理前、处理过程中或处理后拦截请求和响应，比如进行身份的验证
controller		存放接口控制器，处理一些 HTTP 请求，返回响应结果
convert		实体类与视图之间的转换与映射
model		数据模型包，包括 do、dto、vo
	entity	项目中涉及的实体类
	dto	数据传输对象，客户端请求参数的实体封装
	query	接口请求参数封装
	vo	封装接口返回的 VO 视图
mapper		存放 MyBatis 的 mapper 接口，定义数据库操作的方法，每个方法对应一个数据库操作
security		Spring Security 相关代码，子目录在下面介绍
serializer		实体类中序列化工具
service		存放业务逻辑代码的目录
utils		存放一些工具类
	ShareAdminApiApplication	项目启动主类
resources		静态资源目录

（续表）

目录名称	子目录	含义
	log	日志配置，配置日志输出级别、输出格式等
	mapper	mapper 接口类中对应的 xml 格式实现
	application.yml	系统主配置文件
	application-dev.yml	系统在 dev 运行环境下的配置文件，在主配置文件中可以指定运行环境

在这里，我们重点对 security 包的相关内容进行介绍。security 包中的内容如表 2 所示。

表 2　security 包的目录文件内容

包名	含义
cache	存放 token 相关的缓存工具类
config	security 配置，包括密码加密、白名单以及核心配置
exception	定义自定义的 401 异常
filter	过滤器，主要用于用户登录验证
service	重写 security 中的 UserDetailsService，实现自定义登录
user	重写 security 源码中的 UserDetails 实体
utils	secutity 相关工具类

1.2.4　启动项目

开发环境介绍

操作系统：Windows 10 及以上

Web 服务器：Tomcat 10

Java 开发包：JDK 17

数据库：MySQL 8.0

开发工具：IntelliJ IDEA、Visual Studio Code

浏览器：Chrome

1. 将下载好的后台管理系统项目模板进行解压，打开 Idea，点击"Open"，找到刚刚下载解压的后端项目，导入项目（图 2）。

2. 打开设置，修改 Maven 构建工具的配置，框选部分选择自己电脑下的 Maven 相关文件和目录（图 3）。

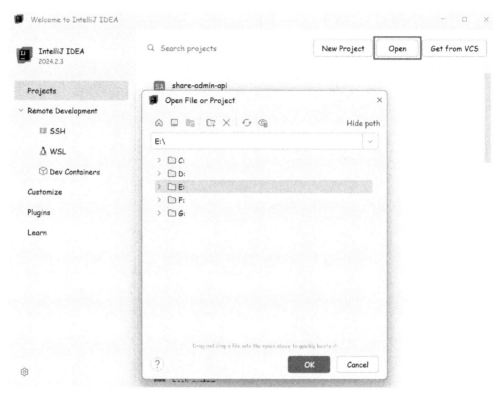

图 2　导入项目

图 3　选择自己电脑下的 Maven 相关文件和目录

3. Maven 依赖加载完成后，找到 resources 目录下的 application-dev.yml 文件，修改 dev 环境下的配置，主要是 MySQL 以及 redis 的配置（图 4）。

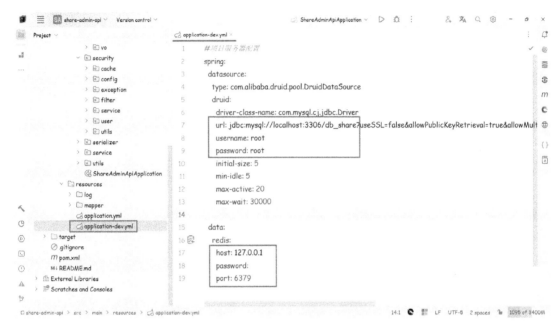

图 4　修改 dev 环境下的配置

4. 修改成自己的配置后，确保 redis 和 MySQL 都正常运行，找到 ShareAdminApiApplication 启动类，准备启动项目（图 5）。

图 5　准备启动项目

运行成功后就可以准备测试接口，接口文档访问地址：http://localhost:8081/share-admin-api/doc.html。

1.2.5 项目功能测试

项目功能测试,也称为行为测试或黑盒测试,是一种软件测试方法,主要关注于系统的实际输出是否符合预期结果,而不关心程序内部的具体实现。功能测试是所有测试工作中最大也是最重要的部分,其主要目的是从用户的角度出发,确保系统的执行与需求一致。测试人员会模拟用户操作,对系统的各项功能进行详细的测试,包括正常情况下的功能验证以及异常情况下的错误处理。

在本小节中,我们通过"认证管理"模块,对项目功能进行简单测试。

1. 成功运行系统后,打开接口文档地址,接口文档账号密码在 application.yml 的配置文件中配置(图6)。

2. 进入文档后,选择请求认证管理中账号密码登录接口,点击调试,按要求输入请求参数,默认账号和密码为 admin/admin,点击"发送",获取 access_token(图7)。

```
# knife4j的增强配置,不需要增强可以不配
knife4j:
  enable: true
  setting:
    language: zh_cn
  basic:
    enable: true    # 开启密码模式
    username: admin # 用户名
    password: 123456 # 密码
```

图 6 配置文件中的初始账号密码

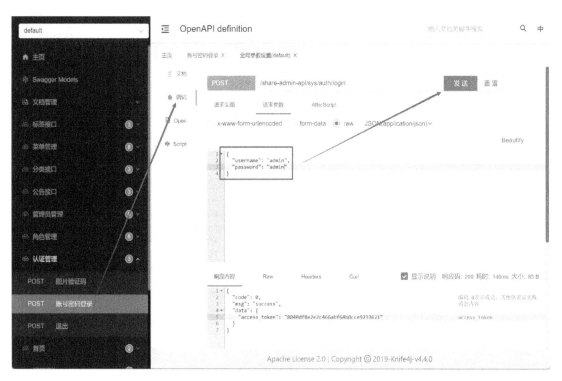

图 7 获取 access_token

3. 在文档管理的全局参数中添加参数,参数名称为 Authorization,参数类型为 header,参数值就用刚刚请求到的 access_token(图 8)。

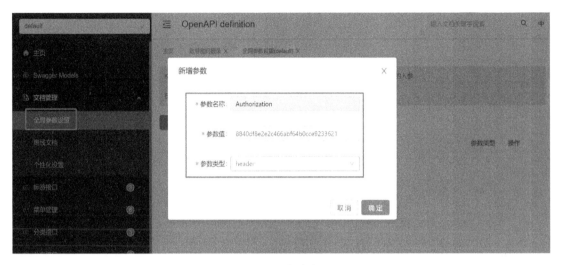

图 8　添加参数

4. 刷新页面,请求【退出登录】接口,会自动带上请求头,测试结果如图 9 所示。

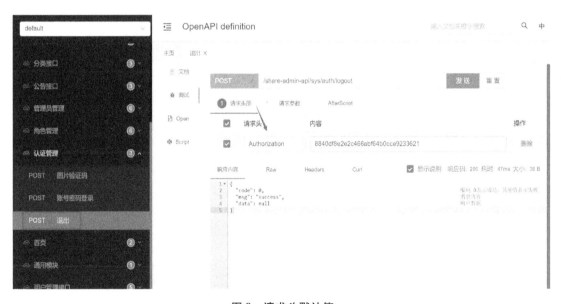

图 9　请求头默认值

Authorization 使用一次就会失效,所以再次请求【退出登录】接口会失败,报 401 未认证错误(图 10)。

图 10　Authorization 失效

2　文件上传

2.1　任务描述

波哥准备为小南和小工详细讲解项目中文件上传功能的实现。

他首先提到,为了确保文件的稳定性和安全性,项目将采用阿里云 OSS 作为云存储服务。因此,首要任务是进行一系列的第三方服务准备工作,包括注册登录阿里云、进行实名认证、创建 AccessKey(AK)和 Bucket,并安装 OSS Browser 来管理 OSS 资源。

完成这些准备工作后,我们需要修改项目的配置文件,将阿里云 OSS 的相关参数加入其中,以便在后续的代码开发中能够正确地与 OSS 服务交互。

接下来,为了实现文件的在线访问,我们需要定义一个 VO(View Object,视图对象)来封装文件的线上链接。在 vo 包下,我们将新建 FileUrlVO 实体类,用于存储和返回文件的 URL 信息。

在 Service 层,我们将实现文件上传的具体业务逻辑。

为了实现文件上传的接口,我们需要在 Controller 层定义相应的接口。

最后,考虑到文件上传功能通常不需要用户登录认证,我们需要在配置文件中放行相应的接口。在 application.yml 中,我们将以 common 开头的接口配置为不需要进行登录校验,以确保用户能够顺利地上传文件。

通过以上的步骤,我们将能够完整地实现文件上传功能,并将文件存储在阿里云 OSS 上,提供稳定的在线访问服务。

2.2 任务分析

2.2.1 第三方服务准备

1. 注册登录阿里云

要使用阿里云的服务,首先需要在阿里云官网注册一个账号并进行登录。你可以通过阿里云官网的顶部导航栏找到"注册"或"登录"的入口,并按照页面提示完成相关操作。

2. 实名认证

注册完账号后,为了保障你的账号安全和服务使用权限,需要进行实名认证。你可以在阿里云账号管理页面中找到实名认证的入口,并按照页面提示完成实名认证流程。

3. 创建 AK

AccessKey(简称"AK")是阿里云用于身份验证和授权访问的凭证。你可以在阿里云 RAM(资源访问管理)控制台中创建 AccessKey(图 11),并确保妥善保管生成的 AccessKey ID 和 AccessKey Secret。可以通过阿里云控制台导航到 RAM 服务,然后找到 AccessKey 管理页面进行操作。

图 11 创建 Accesskey

4. 开通 OSS

阿里云对象存储服务(OSS)是一个海量、安全、低成本、高可靠的云存储服务。在阿里云官网找到 OSS 服务的介绍页面,并根据页面提示开通 OSS 服务。

5. 创建 Bucket

在开通 OSS 服务后,需要创建一个或多个 Bucket 来存储你的数据。Bucket 是 OSS 中存储数据的逻辑单元,具有唯一的名字和地域属性。请注意 Bucket 的命名规则,并选择一个适合业务需求的地域(图 12)。可以在 OSS 控制台中找到 Bucket 管理页面,并按照页面提示创建 Bucket。

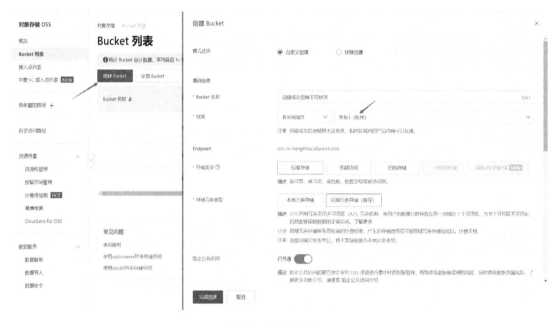

图 12　创建 Bucket

创建成功后,回到 Bucket 列表,选择左侧菜单:权限控制→读写权限(图 13)。

图 13　选择"读写权限"

点击"设置"按钮,选中"公共读写"选项,点击"保存"(图14)。

图14　设置为"公共读写"

在弹出框中点击"继续修改"(图15)。

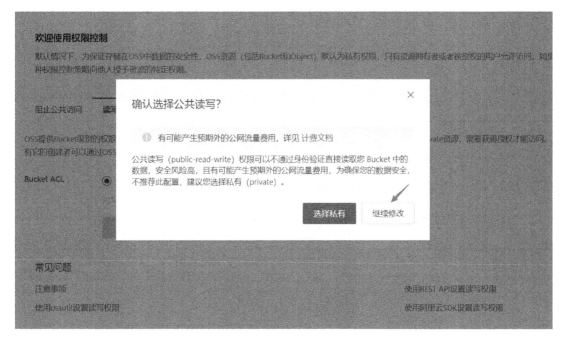

图15　点击"继续修改"

最后点击"保存"(图16)。

图16　点击"保存"

6. OSS Browser

OSS Browser 是阿里云官方提供的 OSS 图形化管理工具,提供类似 Windows 资源管理器的功能。可以在阿里云的产品帮助文档中搜索并查看如何安装和登录 OSS Browser。

下载安装后启动,输入之前申请的 AK 和密钥登录即可(图17)。

图17　AK 登录

2.2.2 修改配置

第三方服务和后台系统项目都准备好之后,在 application.yml 中对应位置可以修改配置。

```yaml
aliyun:
  oss:
    # oss 对外服务的访问域名,按照自己情况修改
    endpoint: oss-cn-hangzhou.aliyuncs.com
    # 访问身份验证中用到用户标识,替换成自己的
    accessKeyId: LTAI5******275JD
    # 用户用于加密签名字符串和 oss 用来验证签名字符串的密钥,替换成自己的
    accessKeySecret: ww529zw******GROuOzc7t
    # oss 的存储空间名
    bucketName: 你的 bucket 名称
```

2.2.3 VO 返回视图封装

项目目录下找到 vo 包,新建 FileUrlVO 实体,用于返回文件的线上链接。

```java
@Data
@Schema(description = "图片 url 上传地址")
@AllArgsConstructor
public class FileUrlVO {

    @Schema(description = "file_url")
    private String fileUrl;

}
```

2.2.4 Service 层业务实现

1. 项目目录下找到 service 包,新建 CommonService 接口,新增 upload 方法,用于上传文件。

```java
public interface CommonService {

    /**
     * 图片上传
     *
     * @param uploadFile
```

```
     * @return
     */
    FileUrlVO upload(MultipartFile uploadFile);

}
```

2. 在 impl 包下新建 CommonServiceImpl 类,实现文件上传方法。首先校验文件格式,设置文件名称,最后调用 AliOSS 服务上传到阿里云 OSS。

```
@Service
public class CommonServiceImpl implements CommonService {
    // 允许上传文件(图片)的格式
    private static final String[] IMAGE_TYPE = new String[]{".bmp", ".jpg",
                                            ".jpeg", ".gif", ".png"};
    @Resource
    private OSSClient ossClient;

    @Value("${aliyun.oss.bucketName}")
    private String bucketName;

    @Override
    public FileUrlVO upload(MultipartFile uploadFile) {

        String returnImgeUrl = "";

        // 校验图片格式
        boolean isLegal = false;
        for (String type : IMAGE_TYPE) {
            if (StringUtils.endsWithIgnoreCase(uploadFile.getOriginalFilename(), type)) {
                isLegal = true;
                break;
            }
        }
        if (!isLegal) {
            // 如果图片格式不合法
            throw new ServerException("图片格式不正确");
        }
```

```java
// 获取文件原名称
String originalFilename = uploadFile.getOriginalFilename();
// 获取文件类型
String fileType = originalFilename.substring(originalFilename.lastIndexOf("."));
// 新文件名称
String newFileName = UUID.randomUUID().toString() + fileType;
// 构建日期路径,例如:OSS 目标文件夹/2020/10/31/文件名
String filePath = new SimpleDateFormat("yyyy/MM/dd").format(new Date());
// 文件上传的路径地址
String uploadImgeUrl = filePath + "/" + newFileName;


// 获取文件输入流
InputStream inputStream = null;
try {
    inputStream = uploadFile.getInputStream();
} catch (IOException e) {
    e.printStackTrace();
}
/**
 * 现在阿里云 OSS 默认图片上传 ContentType 是 image/jpeg
 * 也就是说,获取图片链接后,图片是下载链接,而并非在线浏览链接
 * 因此,这里在上传的时候要解决 ContentType 的问题,将其改为 image/jpg
 */
ObjectMetadata meta = new ObjectMetadata();
meta.setContentType("image/jpg");


//文件上传至阿里云 OSS
ossClient.putObject(bucketName, uploadImgeUrl, inputStream, meta);
// 获取文件上传后的图片返回地址
returnImgeUrl = "https://" + bucketName + "." + ossClient.getEndpoint().getHost() + "/" + uploadImgeUrl;


        return new FileUrlVO(returnImgeUrl);

    }

}
```

2.2.5 Controller 层接口实现

项目目录下找到 controller 包，新建 CommonController 类，定义文件上传接口，调用前文的上传方法。

```
@Tag(name = "通用模块")
@RestController
@RequestMapping("common")
@AllArgsConstructor
public class CommonController {

    private final CommonService commonService;

    @PostMapping(value = "/upload/img")
    @ResponseBody
    @Operation(summary = "图片上传")
    public Result<FileUrlVO> upload(@RequestBody MultipartFile file) {
        return Result.ok(commonService.upload(file));
    }

}
```

2.2.6 接口放行

文件上传理论上不一定需要登录认证，所以在 application.yml 中对接口配置放行，代表 common 开头的全部接口都不需要校验（图 18）。

图 18　设置 common 开头的接口放行

3 后台管理系统前端

3.1 任务描述

在为小南、小工讲解后台管理系统前端的关键开发步骤时,波哥提及了项目启动的首要步骤——下载项目模板。他解释说,为了快速搭建起系统的前端框架,团队将使用预先设计好的项目模板。随后,波哥指导小南和小工如何在 VS Code 中打开项目,并根据项目提供的 README.md 文件,安装必要的插件和依赖,成功启动项目。

项目启动后,波哥强调了动态路由的重要性,并指出为了在页面上显示相应的功能,需要在菜单管理模块中预先创建对应的菜单。他解释说,由于后台管理系统需要管理用户、标签、资源类型、公告以及资源等多个模块,团队需要在现有用户管理的基础上,配置其他几个页面的显示。

波哥进一步指导小南和小工在 src/views 目录下新建 Resources 子目录,并分别创建"资源分类管理""资源管理""公告管理"和"标签管理"等视图文件。随后,他详细说明了如何在"菜单管理"中添加这些新菜单,并强调了在添加菜单时需要注意的事项,如菜单类型、name 属性与 vue 文件名称的对应等。

完成菜单配置后,波哥提到了权限控制的重要性。他解释说,前后端都需要进行权限校验,以确保用户只能访问其被授权的功能。他向小南和小工介绍了项目中约定的权限控制规范,并以"通知管理"模块为例,展示了如何配置菜单项的按钮权限。

最后,波哥提到了字典管理功能,他解释说这一功能可以帮助团队将项目中一些通用的字段进行抽取和集中管理,以提高开发效率和代码的可维护性。他根据前期需求分析和数据库设计,建议团队抽取如"性别""分类类型"等字段作为字典。

通过这一系列的步骤,波哥帮助小南和小工理清了后台管理系统前端开发的主要流程和关键点,为他们的后续工作提供了有力的指导。

3.2 任务分析

3.2.1 下载项目模板

扫描二维码,参考文件夹中"任务三后台管理系统前端项目模板"。

3.2.2 启动项目

将项目通过 VS Code 打开,按照 README.md 文件中推荐的插件,安装插件,安装依赖,运行项目。

```
# 安装依赖
npm install 或 pnpm install

# 本地运行
npm run dev 或 pnpm run dev
```

打开浏览器,输入后台管理系统的登录页地址:http://localhost:4000/,运行效果如图 19 所示。

登录默认账号为:admin,密码为:admin。

图 19　后台管理系统登录页

3.2.3 菜单管理

项目采用的是动态路由,需要我们在菜单管理模块先创建对应的菜单,才可以进行页面显示。

根据前期需求分析,后台管理系统需要对用户、标签、资源类型、公告以及资源进行管理。项目中已经存在用户管理,我们可以在该基础上进行修改,其他几个页面需要进行配置。

3.2.3.1 创建对应的文件目录以及文件

首先,在 src/views 目录下新建 Resource 子目录,创建对应的视图文件,分别为 Category(资源分类管理)、Notice(公告管理)、Resource(资源管理)、Tag(标签管理)。

图 20　创建对应的视图文件

参考下面代码,分别编写几个页面的结构。

<template>

<div>分类管理<div>

<template>

<script setup lang="ts"></script>

<style lang="less" scoped></style>

3.2.3.2 配置对应菜单

在系统管理的"菜单管理"的"新增菜单"中,可以新增动态路由到数据库中。需要注意的是,当菜单类型为"菜单项"时,菜单 name 必须为对应的 vue 文件名称,否则 vue 的 keep-alive 缓存组件对该页面不生效。

首先来配置资源管理模块的根目录,依次选择:系统管理→菜单管理→新增菜单,如图 21 所示。

图21　配置资源管理模块的根目录

刷新页面，即可看到资源管理的一级目录生效（图22）。

图22　一级目录生效

接着在资源管理的一级目录下，分别配置四个菜单项：分类管理、公告管理、标签管理、资源管理。

注意：路由路径可以是小写，但是组件路径和组件名称（图中的Category）一定要跟项目中文件夹的名字一样，菜单项的图标自行选择（图23）。

其他几个配置跟上面一样，配置完成后刷新页面，可以看到刚刚配置的菜单已经正常显示了。因为管理员默认拥有全部菜单，所以不需要额外操作（图24）。

图 23　组件路径

图 24　全部菜单

3.2.3.3　配置对应权限按钮

前后端都需要对应的权限控制，前端配置的授权标识需要跟后端权限标识一样，这样权限才会生效。

在项目中我们约定一套规范：
- view 为查看
- edit 为编辑
- add 为新增
- remove 为删除

- audit 为审批
- export 为导出用户列表
- bonus 为查看用户积分
- ice 为冻结用户

授权标识(多个用逗号分隔,如 sys:menu:list,sys:menu:save),在项目中有几个页面较为特殊,例如资源管理模块,只能进行查看和审批;用户管理模块可以查看、编辑、导出、积分、冻结等。

下面以分类管理模块为例,演示查看权限的配置,如图 25 所示。

图 25　分类管理模块的配置

分类管理的新增、编辑、删除的权限配置如图 26 所示。

图 26　新增、编辑、删除的权限配置

参考完成公告管理、标签管理模块菜单项的按钮权限配置。

资源管理的按钮权限配置如图 27 所示。

图 27　资源管理的按钮权限配置

用户管理的按钮权限配置如图 28 所示。

图 28　用户管理的按钮权限配置

需要使用权限的时候,可以用自定义指令 v-hasPermi 给按钮绑定权限,如下:

```
<!-- 表格操作 -->
<el-button v-hasPermi="['sys:manager:view']" type="primary" link :icon="View" @click="openDrawer
('查看', scope.row)">查看</el-button>
```

```
<el-button v-hasPermi = "['sys:manager:edit']" type = "primary" link : icon = "EditPen" @ click =
"openDrawer('编辑', scope.row)">编辑</el-button>
<el-button v-hasPermi = "['sys:manager:reset-psw']" type = "warning" link : icon = "EditPen" @ click =
"openDrawer('重置', scope.row)">重置密码</el-button>
<el-button v-hasPermi = "['sys:manager:remove']" type = "danger" link : icon = "Delete" @ click =
"deleteAccount(scope.row)">删除</el-button>
```

可以在角色管理里给具体的角色绑定权限,比如新增一个普通用户角色,只有相应模块的查看功能(图29)。

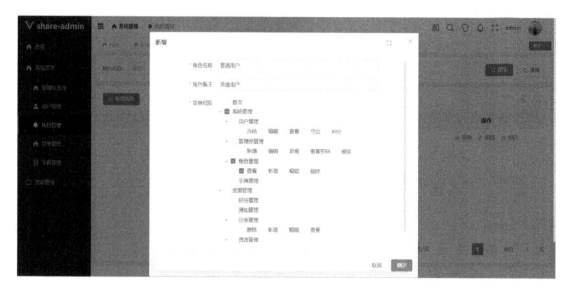

图29 绑定权限

当然,后端也需要对对应的 API 请求做权限校验,以管理员的删除权限举例,给 SysMangerController 的删除(delete)方法一个 PreAuthorize 注解即可。

```
@ PreAuthorize("hasAuthority('sys:manager:remove')")
@ PostMapping("remove")
public Result<String> delete(@ RequestBody List<Integer> idList) {}
```

3.2.4 字典管理

字典管理可以将项目中一些通用字段进行抽取,方便后续使用。例如性别有两个取值:男和女,对应的字典的取值为:0 和 1。

根据分析,结合数据库的设计,我们可以抽取:性别、分类类型、是否置顶、审核状态、是否热门、是否轮播等对应的字段为字典。

图 30　新增性别字典

在系统管理/字典管理中新增字典,比如性别字典,如图 30 所示。然后在操作栏对该字典进行配置,字典的 key 为"名称"(string 类型),value 在"数据值"(int 类型)中,如图 31 所示。

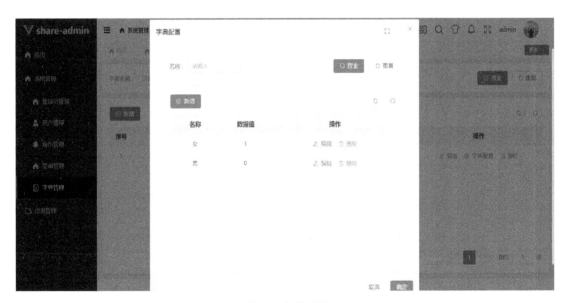

图 31　字典配置

本系统所有字典的展示如图 32 所示。

具体的每个字典的配置名称和取值如表 3 所示。

图 32 系统所有字典列表

表 3 字典名称和取值

字典名称	字典配置
性别	0:男
	1:女
分类类型	0:网盘类型
	1:资源类型
是否置顶	0:否
	1:是
审核状态	0:待审核
	1:审核通过
	2:审核不通过
是否热门	0:否
	1:是
是否轮播	0:否
	1:是

配置好字典后,就可以在页面中使用接口获取字典列表了,例如在管理员管理里添加一列性别,就可以这样来做。

编辑 src/views/System/Manager.vue 文件:

```
import { dictConfigList } from '@/api/modules/dict/dictConfig'

// 表格配置项
const columns: ColumnProps<SysManager.ResManagerList>[] = [
    // 其他列代码
    {
        prop: 'gender',
        label: '性别',
        width: 80,
        enum: () => dictConfigList('gender'),
        fieldNames: { label: 'title', value: 'value' },
        search: { el: 'select', props: { filterable: true } }
    },
]
```

enum 属性定义枚举类型，可以直接写数组，也可以请求接口获取；fieldNames 属性指定 label && value && children 的 key 值（该接口返回的字典 key 字段名叫作 title，所以需要将 label 指向 title）；search 属性指定搜索框的类型。

执行效果如图 33 所示。

图 33　在管理员管理里添加"性别"列

4　任务总结

本次任务涵盖了后台管理系统开发的三个关键小节：后台管理系统后端、文件上传功能实现,以及后台管理系统前端。

后台管理系统后端:在后端开发方面,我们成功搭建了系统的基础架构,完成了数据库的设计和建表工作,并实现了用户认证、权限控制等核心功能。后端服务为前端提供了稳定、可靠的数据支持,确保了整个系统的顺畅运行。

文件上传功能实现:针对文件上传功能,我们采用了阿里云 OSS 作为云存储服务,并完成了相应的准备工作,包括创建 AccessKey、Bucket 等。通过编写后端接口,我们实现了文件的上传、下载、删除等操作,并确保了文件的安全性和稳定性。此外,我们还对上传的文件进行了格式校验和名称处理,提高了系统的健壮性和易用性。

后台管理系统前端:在前端开发方面,我们按照项目模板快速搭建了前端界面,并根据需求完成了菜单管理、页面配置等工作。通过动态路由和权限控制,我们确保了用户只能访问其被授权的页面和功能。同时,我们还创建了字典管理功能,将项目中一些通用的字段进行抽取和集中管理,提高了开发效率和代码的可维护性。最终,我们成功实现了用户、标签、资源类型、公告以及资源等模块的管理功能,为系统的日常运营提供了有力的支持。

任务四

用户管理模块开发

波哥开始分配用户管理模块的开发任务:"小南,你负责开发用户列表展示功能,确保列表能清晰展示用户信息并支持分页和搜索。同时,实现导出用户信息功能,允许用户下载 CSV 或 Excel 文件。小工,你负责开发查看和编辑用户信息的功能,确保用户信息能正确展示和更新。另外,你还需要实现冻结用户账号的功能,用于暂停违规或需要暂停的用户。大家开始工作吧,有问题随时沟通。"

◇ 任务点

- 用户列表展示功能开发;
- 查看、编辑用户信息功能开发;
- 导出用户信息功能开发;
- 冻结用户账号功能开发。

◇ 任务计划

- 任务内容:完成后台管理系统中用户管理模块的开发;
- 任务耗时:预计完成时间为 2~3 h;
- 任务难点:Element Plus 组件的熟练使用。

1 用户列表

1.1 任务描述

波哥召集了小南和小工,准备就后台管理系统中即将上线的用户列表模块进行详细的技术说明和分工。

在后端方面,首先需要准备基础类,包括在 model/query 包下新建 UserQuery 类作为查询实体,用于分页查询并定义相关查询条件;在 model/entity 包下新建 User 实体,对应数据库中的 t_user 表,并在 enums 包下定义用户状态枚举 AccountStatusEnum。同时,为了返回给前端友好的用户信息视图,需要定义 VO,并使用@JsonFormat 注解来格式化日期时间数据。

接下来,需要创建 UserMapper 接口,并在 resource/mapper 文件夹下编写 UserMapper.

xml 文件来实现接口中的方法,特别是分页查询功能。此外,为了方便实体之间的属性复制,还需要在 convert 包下新建 UserConvert 接口,并利用 mapstruct 的@ Mapper 注解实现实体类的转换。

在 Service 层,波哥要求小南新建 UserService 接口并定义分页条件查询方法,同时在 service.impl 包下实现该接口的具体方法,主要调用前面编写的 Mapper 接口方法。

对于前端部分,波哥告诉小工需要定义用户管理模块的相关类型,修改 src/types/dev. d.ts 文件下的用户类型定义,并定义分页查询的接口。

1.2 任务分析

1.2.1 后端基础类准备

1.2.1.1 Query 查询实体创建

model/query 包下新建 UserQuery 类,查询实体主要用于分页查询,继承查询基类。子类中定义了三个查询条件,根据昵称查询、根据手机号查询、根据性别查询。

```
@Data
@EqualsAndHashCode(callSuper = false)
@Schema(description = "用户查询")
public class UserQuery extends Query {
    @Schema(description = "昵称")
    private String nickname;
    @Schema(description = "手机号")
    private String phone;
    @Schema(description = "性别")
    private Integer gender;
}
```

1.2.1.2 PO 实体创建

1. model/entity 包下新建 User 实体,对应数据库表 t_user 表。

```
@Getter
@Setter
@ToString
@TableName("t_user")
public class User {
    @TableId(value = "pk_id", type= IdType.AUTO)
    private Integer pkId;
```

```java
    private String phone;
    private String wxOpenId;
    private String avatar;
    private String nickname;
    private Integer gender;
    private String birthday;
    private Integer bonus;
    private String remark;
    /*
     * @see top.ssy.share.admin.enums.AccountStatusEnum
     */
    private Integer enabled;
    @TableField(value = "delete_flag", fill = FieldFill.INSERT)
    @TableLogic
    private Integer deleteFlag;
    @TableField(value = "update_time", fill = FieldFill.INSERT_UPDATE)
    private LocalDateTime updateTime;
    @TableField(value = "create_time", fill = FieldFill.INSERT)
    private LocalDateTime createTime;
}
```

2. enums 包下新建 AccountStatusEnum,定义用户状态枚举。

```java
@Getter
@AllArgsConstructor
public enum AccountStatusEnum {
    /**
     * 停用
     */
    DISABLE(0, "停用"),
    /**
     * 正常
     */
    ENABLED(1, "正常");

    private final int value;
    private final String name;

    public static String getNameByValue(int value) {
```

```java
            for (AccountStatusEnum s : AccountStatusEnum.values()) {
                if (s.getValue() == value) {
                    return s.getName();
                }
            }
            return "";
        }

        public static Integer getValueByName(String name) {
            for (AccountStatusEnum s : AccountStatusEnum.values()) {
                if (Objects.equals(s.getName(), name)) {
                    return s.getValue();
                }
            }
            return null;
        }
}
```

1.2.1.3　VO 返回视图封装

定义返回给前端的用户信息视图。

@JsonFormat：Jackson 中的注解，用于数据格式化，这里是对 LocalDateTime 进行格式化。

model/vo 包下新建 UserInfoVO 类。

```java
@Data
@Schema(name = "UserInfoVO", description = "用户信息返回 vo")
public class UserInfoVO {
    @Schema(name = "pk_id", description = "用户 id")
    private Integer pkId;
    @Schema(name = "nickname", description = "昵称")
    private String nickname;
    @Schema(name = "phone", description = "手机号")
    private String phone;
    @Schema(name = "wxOpenId", description = "微信 openid")
    private String wxOpenId;
    @Schema(name = "avatar", description = "头像")
```

```
    private String avatar;
    @Schema(name = "gender", description = "性别")
    private Integer gender;
    @Schema(name = "birthday", description = "生日")
    private String birthday;
    @Schema(name = "bonus", description = "积分")
    private Integer bonus;
    @Schema(name = "remark", description = "备注")
    private String remark;
    @Schema(description = "账户状态")
    private Integer enabled;
    @Schema(name = "createTime", description = "创建时间")
    @JsonFormat(pattern = "yyyy-MM-dd HH:mm:ss", timezone = "GMT+8")
    private LocalDateTime createTime;
}
```

1.2.1.4 Mapper 接口

1. mapper 包下新建 UserMapper 接口。同时实现两个方法,分别是按照手机号查询用户和后续会用到的分页条件查询用户。

```
public interface UserMapper extends BaseMapper<User> {

    default User getByPhone(String phone) {
        return this.selectOne(new LambdaQueryWrapper<User>().eq(User::getPhone, phone));
    }

    List<UserInfoVO> getUserPage(Page<UserInfoVO> page, @Param("query") UserQuery query);
}
```

2. resources 文件夹下新建 mapper 文件夹,在 mapper 文件夹下新建 UserMapper.xml 文件,用于实现 UserMapper 接口中的方法,这里只需要实现 getUserPage 方法。

```
<?xml version="1.0" encoding="UTF-8"?>
<!DOCTYPE mapper PUBLIC "-//mybatis.org//DTD Mapper 3.0//EN" "http://mybatis.org/dtd/mybatis-3-mapper.dtd">
<mapper namespace="top.ssy.share.admin.mapper.UserMapper">

    <select id="getUserPage" resultType="top.ssy.share.admin.model.vo.UserInfoVO">
```

```xml
        SELECT tu.* FROM t_user tu
        WHERE tu.delete_flag = 0
        <if test="query.nickname != null and query.nickname != '' ">
            AND tu.nickname LIKE concat('%',#{query.nickname},'%')
        </if>
        <if test="query.phone != null and query.phone != '' ">
            AND tu.phone = #{query.phone}
        </if>
        <if test="query.gender != null">
            AND tu.gender = #{query.gender}
        </if>
        ORDER BY tu.create_time DESC
    </select>

</mapper>
```

1.2.1.5 Convert 实体转换

convert 包下新建 UserConvert,这个接口用作简单的实体属性复制。其中的 @Mapper 注解是 mapstruct 中的注解,它用于帮助我们实现实体类的转换。注意要导入 mapstruct 下的 Mapper 包。

```java
package top.ssy.share.admin.convert;

import org.mapstruct.Mapper;
import org.mapstruct.factory.Mappers;

import top.ssy.share.admin.model.entity.User;
import top.ssy.share.admin.model.vo.UserInfoVO;

import java.util.List;

@Mapper
public interface UserConvert {
    UserConvert INSTANCE = Mappers.getMapper(UserConvert.class);

    List<UserInfoVO> convert(List<User> list);
}
```

我们可以执行命令:mvn clean,然后重新编译项目,到 target 中查看新生成的转换后的实体类，如图 1 所示。

图 1 target 中查看新生成的转换后的实体类

实现类的代码很简单，List 集合的转换底层也是实体类一个一个转换，就是把一个类的属性赋值到另一个类，但是局限性属性名必须一致，否则不会被识别。所以建议使用转换类后检查一下，有没有属性被漏掉，防止参数为空。

1.2.2 Service 层实现

1. service 包下新建 UserService 接口，定义分页条件查询方法。

```
package top.ssy.share.admin.service;

import com.baomidou.mybatisplus.extension.service.IService;
import top.ssy.share.admin.common.result.PageResult;
import top.ssy.share.admin.model.dto.UserEditDTO;
import top.ssy.share.admin.model.entity.User;
import top.ssy.share.admin.model.query.UserQuery;
import top.ssy.share.admin.model.vo.UserInfoVO;


public interface UserService    extends IService<User> {

    PageResult<UserInfoVO> page( UserQuery query );
}
```

2. service.impl 包下新建 UserServiceImpl 实现类，对于接口中的分页方法进行实现，主要是调用前面写的 Mapper 接口方法。

```java
package top.ssy.share.admin.service.impl;

import com.baomidou.mybatisplus.extension.plugins.pagination.Page;
import lombok.AllArgsConstructor;
import lombok.extern.slf4j.Slf4j;
import org.springframework.stereotype.Service;
import top.ssy.share.admin.common.exception.ServerException;
import top.ssy.share.admin.common.model.BaseServiceImpl;
import top.ssy.share.admin.common.result.PageResult;
import top.ssy.share.admin.convert.UserConvert;
import top.ssy.share.admin.enums.AccountStatusEnum;
import top.ssy.share.admin.mapper.UserMapper;
import top.ssy.share.admin.model.entity.User;
import top.ssy.share.admin.model.query.UserQuery;
import top.ssy.share.admin.model.vo.UserInfoVO;
import top.ssy.share.admin.service.UserService;

import java.util.List;


@Slf4j
@Service
@AllArgsConstructor
public class UserServiceImpl extends BaseServiceImpl<UserMapper, User> implements UserService {

    @Override
    public PageResult<UserInfoVO> page(UserQuery query) {
        Page<UserInfoVO> page = new Page<>(query.getPage(), query.getLimit());
        List<UserInfoVO> list = baseMapper.getUserPage(page, query);
        return new PageResult<>(list, page.getTotal());
    }

}
```

1.2.3 Controller 接口实现

controller 包下新建 UserController 类，实现分页查询接口。

```java
package top.ssy.share.admin.controller;

import io.swagger.v3.oas.annotations.Operation;
import io.swagger.v3.oas.annotations.tags.Tag;
import jakarta.validation.Valid;
import lombok.AllArgsConstructor;
import org.springframework.security.access.prepost.PreAuthorize;
import org.springframework.web.bind.annotation.*;
import top.ssy.share.admin.common.result.PageResult;
import top.ssy.share.admin.common.result.Result;
import top.ssy.share.admin.model.query.UserQuery;
import top.ssy.share.admin.model.vo.UserInfoVO;
import top.ssy.share.admin.service.UserService;

@RestController
@AllArgsConstructor
@Tag(name = "用户管理", description = "用户管理")
@RequestMapping("/user")
public class UserController {
    private final UserService userService;

    @PostMapping("/page")
    @Operation(summary = "分页")
    @PreAuthorize("hasAuthority('sys:user:view')")
    public Result<PageResult<UserInfoVO>> page(@RequestBody @Valid UserQuery query) {
        return Result.ok(userService.page(query));
    }

}
```

后端的接口实现完成,后续在管理系统前端页面进行测试。

1.2.4 前端定义类型与定义接口

首先定义用户管理模块的相关类型,为后续开发用户模块提供更好的代码支持,修改 src/types/dev.d.ts 下用户类型定义。

```
/** 用户信息 */
declare interface UserType {
```

```
    pkId: number
    wxOpenId: string
    account: string
    nickname: string
    avatar?: any
    phone: string
    gender: number
    birthday: string
    bonus: number
    remark: string
    deleteFlag: number
    createTime?: string
    updateTime?: string
    enabled: number
}

/** 用户分页请求参数信息 */
declare interface ReqPageUser {
    asc?: boolean
    company?: string
    limit: number
    nickname?: string
    order?: string
    page: number
    phone?: string
}
```

需要添加对应的接口,因为系统默认帮我们提供了查询用户列表、编辑和导出用户的接口。可以在这个基础上复用,只需要单独添加一个冻结用户的接口,需要传入的参数为用户的编号,修改 src/api/modules/user/index.ts 中用户管理模块接口。

```
import http from '@/api'
import { _API } from '@/api/axios/servicePort'

/**
 * @name 用户管理模块
 */
export const UserApi = {
```

```
// 查询用户列表
page: (params: any) => http.post(_API + '/user/page', params),
// 编辑用户
edit: (params: any) => http.post(_API + '/user/edit', params),
// 导出用户列表
export: (params: any) =>
  http.post(_API + '/user/export', params, {
    responseType: 'blob'
  }),
// 冻结用户
freezeUser: (userId: number) => http.post(_API + '/user/enabled? userId=' + userId)
}
```

1.2.5 修改表格页面展示数据

ProTable 组件使用属性透传,支持 el-table && el-table-column 所有属性、事件、方法的调用。

表格组件是基于 Elemet Plus 进行开发,关键词可以搜索 Table,查看对应的属性、事件、方法。

后台管理系统模板对 Element Plus 中 Table 组件进行二次封装,具体属性配置及相关方法,可以搜索关键词 ProTable。

ProTable 的一些主要的列配置项,可直接在 src/components/ProTable/index.vue 文件里看 interface ProTableProps 的属性注释。

ProTable 组件通常需要传递的属性有 ref、colums、requestApi、dataCallback。
- rowKey 属性指定行数据的 key,默认为 id,若不是 id 则需要自己单独配置,如示例所示的 rowKey="pkId",也可以修改 ProTable 组件的 rowKey 默认值进行全局修改。
- 如果暂时没有接口数据,那么可以把属性 requestApi 换成 data,可传入静态 table data 数据。
- 属性 cell-style 来自 element-plus 的 Table 组件的属性,通过组件的属性透传来达到效果。
- columns 为表格配置项,具体用法查看 ProTable 文档。

#tableHeader 为头部自定义插槽,可以放一些头部需要的按钮,也可以自定义配置自己需要的按钮。
#operation="scope" 为表格操作项,可以对表格进行一些操作,例如查看、编辑等,可由实际场景进行配置。

修改 src/views/User/components/UserManage.vue 中用户管理列表相关内容。

```vue
<template>
  <div class="table-box">
    <ProTable
        rowKey="pkId"
        ref="proTable"
        title="用户列表"
        :columns="columns"
        :requestApi="getTableList"
        :initParam="initParam"
        :dataCallback="dataCallback"
        :searchCol="{ xs: 1, sm: 2, md: 3, lg: 4, xl: 4 }"
        :row-style="{ height: '0' }"
        :cell-style="{ padding: '0px' }"
    >
      <!-- 表格操作 -->
      <template #operation="scope">
        <el-button type="primary" link :icon="View" @click="openDrawer('查看', scope.row)" v-hasPermi="['sys:user:view']">查看</el-button>
        <el-button type="primary" link :icon="EditPen" @click="openDrawer('编辑', scope.row)" v-hasPermi="['sys:user:edit']">编辑</el-button>
      </template>
    </ProTable>
    <UserDialog ref="dialogRef" />
  </div>
</template>

<script setup lang="tsx" name="UserManager">
import { ref, reactive } from 'vue'
import { ColumnProps } from '@/components/ProTable/interface'
import ProTable from '@/components/ProTable/index.vue'
import UserDialog from './components/UserDialog.vue'
import { View, EditPen } from '@element-plus/icons-vue'
import UserApi from '@/api/modules/user'
import { dictConfigList } from '@/api/modules/dict/dictConfig'
// 获取 ProTable 元素，调用其获取刷新数据方法（还能获取当前查询参数,方便导出携带参数）
const proTable = ref()
```

```js
// 如果表格需要初始化请求参数，直接定义传给 ProTable(之后每次请求都会自动带上该参数,此参数
// 更改之后也会一直带上,改变此参数会自动刷新表格数据)
const initParam = reactive({})


// dataCallback 是对于返回的表格数据做处理,如果你后台返回的数据不是 list && total 这些字段,那么
// 你可以在这里处理成这些字段
const dataCallback = (data: any) => {
  return {
    list: data.list,
    total: data.total
  }
}


// 默认不做操作就直接在 ProTable 组件上绑定 :requestApi="getUserList"
const getTableList = (params: any) => {
  return UserApi.page(params)
}


// 表格配置项
const columns: ColumnProps<UserType>[] = [
  { type: 'selection', fixed: 'left', width: 60 },
  {
    prop: 'avatar',
    label: '头像',
    width: 70,
    // 自定义渲染,设置头像在表格内部居中
    render: (scope) => {
      return (
        <div class={['flex', 'justify-center', 'p-1']}>
          <el-avatar shape={'square'} size={30} src={scope.row.avatar} />
        </div>
      )
    }
  },
  {
    prop: 'nickname',
    showOverflowTooltip: true,
    label: '用户名',
```

```
        width: 100,
        search: {
          el: 'input',
          props: { placeholder: '请输入用户名' }
        }
      },
      {
        prop: 'phone',
        label: '手机号',
        search: {
          el: 'input',
          props: { placeholder: '请输入手机号' }
        },
        width: 120
      },
      { prop: 'bonus', label: '积分', width: 120, sortable: true },
      {
        prop: 'gender',
        label: '性别',

        width: 100,
        // enum: [
        //   {
        //     title: '男',
        //     value: 0
        //   },
        //   {
        //     title: '女',
        //     value: 1
        //   }
        // ],
        // 使用枚举
        enum: () => dictConfigList('gender'),
        // 配置头部搜索
        search: {
          el: 'select',
          props: {
            filterable: true,
```

```
      placeholder: '请选择性别'
    }
  },
  // 配置展示字段属性
  fieldNames: { label: 'title', value: 'value' },
  render: (scope) => {
    // 自定义返回不同类型
    return <el-tag type={scope.row.gender === 0 ? 'success' : 'warning'}>{scope.row.gender === 0 ? '男' : '女'}</el-tag>
  }
},
{
  prop: 'birthday',
  label: '生日'
},
{
  prop: 'enabled',
  label: '状态',
  width: 100,
  render: (scope) => {
    // 自定义返回不同类型
    return <el-tag type={scope.row.enabled === 0 ? 'warning' : 'primary'}>{scope.row.enabled === 0 ? '禁用' : '启用'}</el-tag>
  }
},
{
  prop: 'createTime',
  label: '创建时间',
  width: 200
},
{ prop: 'operation', label: '操作', fixed: 'right', width: 340 }
]

// 打开 drawer(新增、查看、编辑)
const dialogRef = ref()
const openDrawer = (title: string, row: Partial<UserType> = {}) => {}
</script>
```

修改表格页面展示数据效果如图 2 所示。

图 2　页面效果

2　查看和编辑用户信息

2.1　任务描述

小南和小工准备实现查看和编辑用户信息功能。

波哥首先向小南和小工介绍了用户列表功能的需求和整体设计。为了支持用户信息的修改，后端需要封装一个 DTO（数据传输对象）来接收前端传递的修改数据。波哥指示小南在 dto 包下新建 UserEditDTO 类，用于封装修改用户信息所需的字段，并在 UserConvert 类中新增转换方法，以便在 Service 层进行实体与 DTO 之间的转换。

波哥告诉小南，在 UserService 接口中需要新增一个更新用户信息的方法。随后，小南需要在 service.impl 包下新建 UserServiceImpl 实现类，并实现接口中定义的更新用户信息的方法。这个方法将调用 Mapper 层提供的数据访问功能，完成用户信息的更新操作。

波哥又转向 Controller 层的实现。他解释说，Controller 层将负责接收前端的请求，调用 Service 层提供的业务逻辑方法，并将处理结果返回给前端。小工需要确保 Controller 层能够正确地处理前端发送的 HTTP 请求，并调用相应的 Service 层方法。

通过这次讨论，小南和小工对查看和编辑用户信息功能的实现有了清晰的认识，并准备开始各自的开发工作。

2.2　任务分析

2.2.1　DTO 请求实体封装

1. dto 包新建 UserEditDTO，用于修改用户信息。

@Data
@Schema(description = "用户修改 dto")

```
public class UserEditDTO implements Serializable {

    @Serial
    private static final long serialVersionUID = -3975140673973740199L;

    @NotNull
    @Schema(description = "用户id")
    private Integer pkId;
    @Schema(description = "用户名")
    private String nickname;
    @Length(min = 11, max = 11, message = "手机号长度为11位")
    @Schema(description = "手机号")
    private String phone;
    @Schema(description = "性别")
    private Integer gender;
    @Schema(description = "头像")
    private String avatar;
    @Schema(description = "生日")
    private String birthday;
}
```

2. 新增转换方法。

```
package top.ssy.share.admin.convert;

import org.mapstruct.Mapper;
import org.mapstruct.factory.Mappers;
import top.ssy.share.admin.model.dto.UserEditDTO;
import top.ssy.share.admin.model.entity.User;
import top.ssy.share.admin.model.vo.UserInfoVO;

import java.util.List;

@Mapper
public interface UserConvert {
    UserConvert INSTANCE = Mappers.getMapper(UserConvert.class);

    User convert(UserEditDTO dto);

    List<UserInfoVO> convert(List<User> list);
}
```

2.2.2 Service 层实现

1. UserService 接口,更新用户信息。

```
void update(UserEditDTO dto);
```

2. service.impl 包下新建 UserServiceImpl 实现类,实现接口中的三种方法。

```
@Override
public void update(UserEditDTO dto) {
    User user = UserConvert.INSTANCE.convert(dto);
    User byPhone = baseMapper.getByPhone(user.getPhone());
    if (byPhone != null && !byPhone.getPkId().equals(user.getPkId())) {
        throw new ServerException("手机号已存在");
    }
    baseMapper.updateById(user);
}
```

2.2.3 Controller 接口实现

UserController 类实现更新接口。

```
@PostMapping("edit")
@Operation(summary = "修改")
@PreAuthorize("hasAuthority('sys:user:edit')")
public Result<String> update(@RequestBody @Valid UserEditDTO dto) {

    userService.update(dto);

    return Result.ok();
}
```

后端的接口实现完成,后续在管理系统前端页面进行测试。

2.2.4 打开弹出层传递数据

当打开弹出层时,需要传递标题以及相关数据。根据不同的标题,决定表单的禁用状态。

在查看模式下,将全部表单禁用,以防用户对表单内容进行修改。这可以确保数据的完整性和稳定性。

而在编辑模式下,表单处于可编辑状态,用户可以修改表单内容。这种设计有助于精确控制用户对数据的变更操作。

在 src/views/User/components/UserManage.vue 中,配置弹框方法传递对应数据。

```tsx
<script setup lang="tsx" name="UserManager">
// 打开 drawer(新增、查看、编辑)
const dialogRef = ref()
const openDrawer = (title: string, row: Partial<UserType> = {}) => {
  let params = {
    // 标题
    title,
    // 当前行
    row: { ...row },
    // 是否可以查看
    isView: title === '查看',
    // 需要调用的方法
    api: title === '编辑' ? UserApi.edit : '',
    // 修改后重新刷新表格列表
    getTableList: proTable.value.getTableList,
    // 弹出层最大高度
    maxHeight: '400px'
  }
  // 调用子组件打开弹框方法
  switch (title) {
    case '查看':
      dialogRef.value.acceptParams(params)
      break
    case '编辑':
      dialogRef.value.acceptParams(params)
      break
    default:
      break
  }
}
</script>
```

2.2.5 修改弹出层表单

编辑页面是一个用于展示和编辑用户相关信息的界面。在这个页面中,主要展示一个表单,其中包含了用户的各种信息。

该表单的一个重要功能是展示用户头像上传的选项。用户可以通过点击该选项,上传自己的头像图片。

除了头像上传功能外,每个表单项都有其自身的校验规则。这些规则用于确保输入的信息符合特定的格式、逻辑或业务要求。

关于单选组件、表单组件、日期组件、输入框组件的相关属性及方法,可以查看 Element Plus 官方文档进行了解。

修改组件 src/views/User/components/UserDialog.vue,修改弹出框中对应的表单配置为我们所需要的。

> 我们需要对时间默认值进行处理,使用 date-picker 组件返回的默认值是 Date 类型。而后端需要存放字符串类型,这时需要前端进行处理,我们需要获取选择日期通过字符串进行拼接,最终在提交时重新赋值给生日变量。

```
<template>
  <Dialog :model-value="dialogVisible" :title="dialogProps.title" :fullscreen="dialogProps.fullscreen" :max-height="dialogProps.maxHeight" :cancel-dialog="cancelDialog">
    <div :style="'width: calc(100% - ' + dialogProps.labelWidth! / 2 + 'px)'">
      <el-form
        ref="ruleFormRef"
        label-position="right"
        :label-width="dialogProps.labelWidth + 'px'"
        :rules="rules"
        :model="dialogProps.row"
        :disabled="dialogProps.isView"
        :hide-required-asterisk="dialogProps.isView"
      >
        <el-form-item label="用户名" prop="nickname">
          <el-input v-model="dialogProps.row!.nickname" placeholder="" clearable></el-input>
        </el-form-item>
        <el-form-item label="用户头像" prop="avatar">
          <UploadImg v-model:image-url="dialogProps.row!.avatar" width="135px" height="135px" :file-size="5">
            <template #empty>
              <el-icon><Avatar /></el-icon>
              <span>请上传头像</span>
            </template>
            <template #tip>头像大小不能超过 5M </template>
          </UploadImg>
```

```
            </el-form-item>
            <el-form-item label="手机号" prop="phone">
                <el-input v-model="dialogProps.row!.phone" placeholder="" clearable></el-input>
            </el-form-item>
            <el-form-item label="性别" prop="gender">
                <el-radio-group v-model="dialogProps.row!.gender">
                    <el-radio :label="0" border>男</el-radio>
                    <el-radio :label="1" border>女</el-radio>
                </el-radio-group>
            </el-form-item>
            <el-form-item label="生日" prop="birthday">
                <el-date-picker v-model="dialogProps.row!.birthday" type="date" placeholder="选择日期" clearable></el-date-picker>
            </el-form-item>
        </el-form>
    </div>
    <template #footer>
        <slot name="footer">
            <el-button @click="cancelDialog">取消</el-button>
            <el-button type="primary" v-show="!dialogProps.isView" @click="handleSubmit">确定</el-button>
        </slot>
    </template>
    </Dialog>
</template>

<script setup lang="ts">
import { ref, reactive } from 'vue'
import { ElMessage, FormInstance } from 'element-plus'
import { Dialog } from '@/components/Dialog'
import UploadImg from '@/components/Upload/Img.vue'
// 定义对应的类型
interface DialogProps {
    title: string
    isView: boolean
    fullscreen?: boolean
    row: any
```

```
  labelWidth?: number
  maxHeight?: number | string
  api?: (params: any) => Promise<any>
  getTableList?: () => Promise<any>
}

// 是否展示弹框
const dialogVisible = ref(false)
// 定义相关属性,用于接收
const dialogProps = ref<DialogProps>({
  isView: false,
  title: '',
  row: {},
  labelWidth: 160,
  fullscreen: true,
  maxHeight: '500px'
})

// 接收父组件传过来的参数
const acceptParams = (params: DialogProps): void => {
  // 1.合并参数
  params.row = { ...dialogProps.value.row, ...params.row }
  // 2.合并重新赋值
  dialogProps.value = { ...dialogProps.value, ...params }
  // 3.打开弹框
  dialogVisible.value = true
}

// 暴露给父组件的事件
defineExpose({
  acceptParams
})
// 校验规则
const rules = reactive({
  nickname: [
    { required: true, message: '请输入账号', trigger: 'blur' },
    {
      min: 2,
      max: 10,
```

```
        message: '长度在 2 到 10 个字符',
        trigger: 'blur'
      }
    ],
    phone: [
      { required: true, message: '请输入手机号', trigger: 'blur' },
      // 自定义校验规则
      {
        pattern: /^1[3-9]\d{9}$/,
        message: '请输入正确的手机号',
        trigger: 'blur'
      }
    ],
    avatar: [{ required: true, message: '请上传头像', trigger: 'blur' }],
    gender: [
      {
        required: true,
        message: '请选择性别',
        trigger: 'change'
      }
    ],
    birthday: [
      {
        required: true,
        message: '请选择生日',
        trigger: 'change'
      }
    ]
})
// 表单组件实例
const ruleFormRef = ref<FormInstance>()
// 提交事件
const handleSubmit = () => {
  // 进行表单校验,需要通过才能进行下面的操作
  ruleFormRef.value!.validate(async (valid) => {
    if (!valid) return
    try {
```

```
      // 将生日从 Date 类型转换为字符串
      const date = new Date(dialogProps.value.row.birthday)
      const year = date.getFullYear()
      const month = String(date.getMonth() + 1).padStart(2, '0')
      const day = String(date.getDate()).padStart(2, '0')
      const formattedDate = `${year}-${month}-${day}`
      dialogProps.value.row.birthday = formattedDate
      // 调用保存方法
      await dialogProps.value.api!(dialogProps.value.row)
      ElMessage.success({ message: `${dialogProps.value.title}成功！` })
      // 重新获取表单数据
      dialogProps.value.getTableList!()
      // 关闭弹框
      dialogVisible.value = false
      // 重置表单输入内容
      ruleFormRef.value!.resetFields()
      // 清除内容
      cancelDialog(true)
    } catch (error) {
      console.log(error)
    }
  })
}
// 删除展示内容,根据传递参数决定是否删除
const cancelDialog = (isClean?: boolean) => {
  dialogVisible.value = false
  let condition = ['查看', '编辑']
  if (condition.includes(dialogProps.value.title) || isClean) {
    dialogProps.value.row = {}
    ruleFormRef.value!.resetFields()
  }
}
</script>

<style scoped lang="less"></style>
```

修改用户信息页面效果如图 3 所示。

图 3　修改用户信息页面效果图

3　导出用户

3.1　任务描述

由于导出功能对于用户管理模块来说至关重要,波哥决定使用流行的 EasyExcel 工具来简化实现过程,并建议小南和小工自行搜索 EasyExcel 工具的使用方法。

波哥提到,为了使用 EasyExcel,需要在系统的 pom.xml 文件中添加相应的依赖。他指示小南在 properties 和 dependencies 标签下分别添加所需的配置和依赖项。

接下来,波哥提到需要为导出的实体类添加注解。他解释说,之前创建的 UserInfoVO 类将作为导出的数据源,因此需要在该类的每个字段上添加合适的注解。@ExcelIgnore 注解用于标记不需要导出的字段,而@ExcelProperty 注解则用于指定需要导出字段的列名,并通过 converter 属性指定导出时的转换类。

为了处理特殊字段的导出,如性别和用户账号状态,需要在 converter 包下新建两个转换类:GenderConverter 用于转换性别字段,UserEnabledConverter 用于转换用户账号状态字段。这些转换类将确保导出的数据在 Excel 中能够以用户易于理解的形式呈现。

在 Service 实现方面,UserService 接口需要新增一个导出方法。由于导出操作是异步的,并且需要在方法内部设置返回内容,因此该方法应定义为 void 类型。波哥建议小南参考 EasyExcel 的官方教程来实现 UserServiceImpl 中的导出方法。

在 Controller 层,需要在 UserController 中添加一个新的数据导出接口,以便前端能够触发导出操作。

最后,则是前端配置导出按钮的部分。在用户管理界面中需要添加一个导出按钮,并配置相应的图标、权限和点击事件。点击按钮后,应弹出一个提示框以确认用户是否要进行导出操作。一旦用户确认，将调用封装的 useDownload 方法进行导出。波哥建议小工参考 Element Plus 的官方文档来了解 ElMessageBox 组件的属性和方法，以便实现所需的弹出框功能。

3.2 任务分析

3.2.1 添加依赖

在 pom.xml 文件下添加 EasyExcel 的版本声明和依赖引入，分别在 properties 和 dependencies 标签下，增加以下代码

```xml
<easyexcel.version>3.3.3</easyexcel.version>
```

```xml
<dependency>
    <groupId>com.alibaba</groupId>
    <artifactId>easyexcel</artifactId>
    <version>${easyexcel.version}</version>
</dependency>
```

3.2.2 导出实体添加注解

找到之前创建的 UserInfoVO 类，在每个字段上新增注解。如下：

@ExcelIgnore：代表这个字段不需要导出到 Excel 表格；
@ExcelProperty(value = "性别", converter = GenderConverter.class)：字段需要导出，value 指定了导出的列名，converter 指定导出的转换类。

```java
@Data
@Schema(name = "UserInfoVO", description = "用户信息返回 vo")
public class UserInfoVO {
    @ExcelIgnore
    @Schema(name = "pk_id", description = "用户 id")
    private Integer pkId;
    @ExcelProperty("昵称")
    @Schema(name = "nickname", description = "昵称")
    private String nickname;
    @ExcelProperty("手机号")
    @Schema(name = "phone", description = "手机号")
    private String phone;
    @ExcelProperty("微信 openid")
```

@Schema(name = "wxOpenId", description = "微信 openid")
private String wxOpenId;
@ExcelProperty("头像")
@Schema(name = "avatar", description = "头像")
private String avatar;
// TODO excel 导出性别转换
@ExcelProperty(value = "性别", converter = GenderConverter.class)
@Schema(name = "gender", description = "性别")
private Integer gender;
@ExcelProperty("生日")
@Schema(name = "birthday", description = "生日")
private String birthday;
@ExcelProperty("积分")
@Schema(name = "bonus", description = "积分")
private Integer bonus;
@ExcelProperty("备注")
@Schema(name = "remark", description = "备注")
private String remark;
@ExcelProperty(value = "账户状态", converter = UserEnabledConverter.class)
@Schema(description = "账户状态")
private Integer enabled;
@ExcelProperty("创建时间")
@Schema(name = "createTime", description = "创建时间")
@JsonFormat(pattern = "yyyy-MM-dd HH:mm:ss", timezone = "GMT+8")
private LocalDateTime createTime;
}
```

### 3.2.3　Converter 导出参数转换类实现

某些参数导出可能需要一些额外操作,例如这里的性别是一个 Integer 值,但是存储在 Excel 中肯定需要能看懂的中文,所以需要自定义一个转换类去实现。

1. converter 包下新建 GenderConverter 类,用于性别的转换。

```
package top.ssy.share.admin.convert;

import com.alibaba.excel.converters.Converter;
import com.alibaba.excel.converters.WriteConverterContext;
import com.alibaba.excel.enums.CellDataTypeEnum;
import com.alibaba.excel.metadata.data.WriteCellData;
```
```

```java
public class GenderConverter implements Converter<Integer> {
    @Override
    public Class<?> supportJavaTypeKey() {
        return String.class;
    }

    @Override
    public CellDataTypeEnum supportExcelTypeKey() {
        return CellDataTypeEnum.STRING;
    }

    @Override
    public WriteCellData<?> convertToExcelData(WriteConverterContext<Integer> context) {
        Integer value = context.getValue();
        if (value == 0) {
            return new WriteCellData<>("男");
        } else if (value == 1) {
            return new WriteCellData<>("女");
        } else {
            return new WriteCellData<>("未知");
        }
    }
}
```

2. 新建 UserEnabledConverter，转换用户账号状态。

```java
package top.ssy.share.admin.convert;

import com.alibaba.excel.converters.Converter;
import com.alibaba.excel.converters.WriteConverterContext;
import com.alibaba.excel.enums.CellDataTypeEnum;
import com.alibaba.excel.metadata.data.WriteCellData;
import top.ssy.share.admin.enums.AccountStatusEnum;


public class UserEnabledConverter implements Converter<Integer> {
    @Override
    public Class<?> supportJavaTypeKey() {
        return String.class;
```

}

   @Override
   public CellDataTypeEnum supportExcelTypeKey() {
       return CellDataTypeEnum.STRING;
   }

   @Override
   public WriteCellData<?> convertToExcelData(WriteConverterContext<Integer> context) {
       return new WriteCellData<>(AccountStatusEnum.getNameByValue(context.getValue()));
   }
}

3.2.4 Service 层实现

1. UserService 接口新增导出方法,必须使用 void,因为在方法内部会设置返回内容。

```
void export(UserQuery query, HttpServletResponse response);
```

2. UserServiceImpl 实现导出方法,在官网有教程,可以搜索参考。

```
@Override
public void export(UserQuery query, HttpServletResponse response) {
    LambdaQueryWrapper<User> wrapper = new LambdaQueryWrapper<>();
    wrapper.like(StringUtils.isNotBlank(query.getNickname()), User::getNickname, "%" + query.getNickname() + "%")
            .eq(StringUtils.isNotBlank(query.getPhone()), User::getPhone, query.getPhone())
            .eq(query.getGender() != null, User::getGender, query.getGender());
    List<User> list = baseMapper.selectList(wrapper);
    List<UserInfoVO> excelData = UserConvert.INSTANCE.convert(list);
    try {
        String fileName = URLEncoder
                .encode("用户信息" + System.currentTimeMillis() + ".xls", StandardCharsets.UTF_8)
                .replaceAll("\\+", "%20");
        response.setContentType("application/vnd.openxmlformats-officedocument.spreadsheetml.sheet;charset=UTF-8");
        response.setCharacterEncoding(StandardCharsets.UTF_8.toString());
        response.setHeader("Access-Control-Expose-Headers", "Content-Disposition");
        response.setHeader("Content-Disposition", "attachment;filename=" + fileName);
```

```
            EasyExcelFactory.write(response.getOutputStream(), UserInfoVO.class)
                    .charset(StandardCharsets.UTF_8)
                    .excelType(ExcelTypeEnum.XLS)
                    .sheet()
                    .doWrite(excelData);
        } catch (Exception e) {
            log.error("导出用户信息异常", e);
            throw new ServerException("导出用户信息异常");
        }
    }
}
```

3.2.5 Controller 层接口实现

UserController 中新增数据导出接口,参考代码如下:

```
@PostMapping("export")
@Operation(summary = "导出")
@PreAuthorize("hasAuthority('sys:user:export')")
public void export(@RequestBody UserQuery query, HttpServletResponse response) {
    userService.export(query, response);
}
```

3.2.6 配置导出按钮进行导出

在用户管理界面中,需要增加一个用户导出功能。由于该功能是基于开源框架实现的,因此无需我们进行额外的配置工作,用法参考下方代码,直接调用封装的 hooks 即可。

需要先在表格头部添加一个导出的按钮,配置对应的图标和权限、绑定对应的点击事件即可,导出方法为直接导出 useDownload 方法调用即可。

点击按钮后需要弹出提示框,通过后才能进行导出,需要传递导出的方法、标题及当前列表数据。具体弹出框相关属性及方法,可以查看 Element Plus 官网中 ElMessageBox。

```
<template>
  <div class="table-box">
    <ProTable
    >
      <!-- 表格 header 按钮 -->
      <template #tableHeader>
        <el-button type="primary" :icon="Download" plain @click="downloadFile" v-hasPermi="['sys:user:export']">导出用户</el-button>
      </template>
      <!-- 表格操作 -->
```

```
            <template #operation="scope">
            </template>
        </ProTable>
    </div>
</template>

<script setup lang="tsx" name="UserManager">
import { Download } from '@element-plus/icons-vue'
import { useDownload } from '@/hooks/useDownload'
import { ElMessageBox } from 'element-plus'
// 导出用户列表
const downloadFile = async () => {
    ElMessageBox.confirm('确认导出用户数据？', '温馨提示', { type: 'warning' }).then(() =>
useDownload(UserApi.export, '用户列表', proTable.value?.searchParam))
}
</script>
```

点击"导出用户"即可打开提示框框,点击"确定"按钮即可导出用户列表,页面效果如图4所示。

图4　导出用户数据确认图

4　冻结用户

4.1　任务描述

冻结用户是用户管理模块的功能增强。这个模块需要实现用户账号的冻结和解冻功能,以确保系统安全并满足业务需求。

波哥首先向小南和小工解释了 DTO(数据传输对象)在请求实体封装中的重要性。他指导小南新建一个名为 UserEnabledDTO 的类,这个类将用于封装冻结或解冻用户时所需的必要信息。小南迅速编写好了代码,明确了冻结/解冻请求的参数。

在 Service 层的实现部分,需要在 UserService 接口中新增两个方法:freezeUser 和 unfreezeUser,用于处理用户账号的冻结和解冻逻辑。小工在 UserServiceImpl 类中实现了这两个方法,它们通过修改数据库中用户账号的状态字段来实现冻结和解冻操作。

在 Controller 层的实现部分,需要在 UserController 中添加新的接口,用于接收前端发送的冻结/解冻请求,并调用 Service 层的方法进行处理。

在前端界面开发的部分,波哥向小南和小工展示了用户管理页面,并解释了如何配置冻结及启用按钮。他强调,按钮的显示状态应该根据用户当前的状态来判断,冻结和启用状态需要通过属性来判断。他还演示了如何绑定函数并传入标题和当前行数据,以便在用户点击按钮时触发相应的处理函数。

4.2 任务分析

4.2.1 DTO 请求实体封装

新建 UserEnabledDTO,用于冻结/解冻用户。

```
@Data
@Schema(description = "用户状态修改 dto")
public class UserEnabledDTO {
    @Schema(description = "用户 id")
    private Integer pkId;
    @Schema(description = "账户状态")
    private Integer enabled;
}
```

4.2.2 Service 层实现

1. UserService 新增冻结/解冻用户方法。

```
void enabled(Integer userId);
```

2. UserServiceImpl 实现方法本质上就是修改用户账号状态。

```
@Override
public void enabled(Integer userId) {
    User user = baseMapper.selectById(userId);
```

```java
    if (user == null) {
        throw new ServerException("用户不存在");
    }
    user.setEnabled(user.getEnabled() == AccountStatusEnum.ENABLED.getValue() ? 0 : 1);
    baseMapper.updateById(user);
}
```

4.2.3 Controller 层实现

UserController 新增账户状态修改接口。

```java
@PostMapping("enabled")
@Operation(summary = "账户状态修改")
@PreAuthorize("hasAuthority('sys:user:ice')")
public Result<String> enabled(@RequestParam Integer userId) {
    userService.enabled(userId);
    return Result.ok();
}
```

4.2.4 前端点击按钮进行冻结事件处理

是否冻结用户根据用户状态进行判断，冻结和启用则需要通过属性来进行判断，提前定义好对应按钮，我们需要绑定好函数之后传入标题和当前行。

点击按钮后弹出提示框，当用户选择"确定"时，传入用户的编号，完成后需要重新刷新表格，重新获取数据。当用户被冻结时，该账户就不能在客户端登录了。

在用户管理页面，配置好冻结及启用按钮，绑定对应事件传递当前行内容即可。

```html
<template>
    ...
    <!-- 表格操作 -->
    <template #operation="scope">
        <el-button type="primary" link :icon="View" @click="openDrawer('查看', scope.row)" v-hasPermi="['sys:user:view']">查看</el-button>
        <el-button type="primary" link :icon="EditPen" @click="openDrawer('编辑', scope.row)" v-hasPermi="['sys:user:edit']">编辑</el-button>
        <el-button type="primary" link :icon="Key" @click="actionUser('冻结', scope.row)" v-if="scope.row.enabled === 1" v-hasPermi="['sys:user:ice']">冻结用户</el-button>
        <el-button type="primary" link :icon="Key" @click="actionUser('启用', scope.row)" v-else v-hasPermi="['sys:user:ice']">启用用户</el-button>
```

```
      </template>
</template>

<script setup lang="tsx" name="UserManager">
import { Key } from '@element-plus/icons-vue'
import { ElMessage, ElMessageBox } from 'element-plus'

// 冻结用户
const actionUser = async (title: string, row: UserType) => {
  ElMessageBox.confirm(`确认${title}【${row.nickname}】用户?`, '温馨提示', { type: 'warning'
}).then(async () => {
    try {
      // 调用冻结函数
      await UserApi.freezeUser(row.pkId)
      ElMessage.success(`${title}用户成功`)
      // 重新获取数据
      proTable.value.getTableList()
    } catch (error) {
      ElMessage.error(`${title}用户失败`)
    }
  })
}
</script>
```

冻结用户效果如图5所示,首先弹出提示框选择是否冻结,确认后,查看到页面用户状态发生对应的改变(图6)。

图5 冻结用户确认图

图 6　用户状态冻结成功效果图

5　任务总结

本次任务主要完成了后台管理系统中用户管理模块的开发,包括用户列表展示、查看和编辑用户信息、导出用户信息、冻结用户账号等,并做了接口测试和前后端联调。

在用户列表展示方面,我们设计并实现了一个清晰易用的界面,能够直观地展示所有用户的基本信息。通过分页和搜索功能,管理员可以快速定位到需要关注的用户,提高了工作效率。

查看和编辑用户信息功能允许管理员对用户的详细信息进行查看和编辑。无论是用户的个人资料、联系方式还是角色权限,都可以在这一功能中进行管理和维护。这一功能的实现,确保了用户数据的准确性和完整性,为系统的正常运行提供了有力保障。

在导出用户数据方面,管理员可以通过简单的操作,将用户数据导出到本地,方便进行后续的数据分析和处理。这一功能的实现,不仅提高了数据的可移植性,也为管理员提供了更多的工作便利。

在冻结用户账号功能中,我们实现了对用户账号状态的灵活控制。当管理员发现某个用户存在异常行为或安全隐患时,可以迅速通过冻结账号来防止潜在风险。同时,我们也为管理员提供了解冻账号的功能,以便在确认用户身份后恢复其正常权限。这一功能的实现,不仅增强了系统的安全性,也体现了对用户权益的尊重和保护。

任务五

积分管理模块开发

波哥在团队会议上对小南和小工说:"接下来,我们需要开发积分管理模块。这个模块将包含用户积分的展示,并且会涉及父子组件的交互。你们有没有信心按时完成?"

小南迅速回应:"没问题,波哥。我们可以先设计好后端接口,确保数据的准确性。然后再进行前端组件的开发和交互设计。"

小工点头表示同意:"对,前后端要紧密配合。我们可以先讨论一下后端接口的实现细节,然后同步给前端进行开发。"

大家迅速进入工作状态,开始着手积分管理模块的开发。

◇ 任务点

- 积分管理模块的后端实现;
- 积分管理模块的前端实现。

◇ 任务计划

- 任务内容:完成后台管理系统中积分管理模块的开发;
- 任务耗时:预计完成时间为 30 min~1 h;
- 任务难点:父子组件交互。

1 后端基础类创建

1.1 任务描述

积分系统是近期项目中重要的一部分,它将为用户行为提供奖励,激励用户更加活跃地参与平台活动。

波哥建议小南和小工首先实现后端的基础类,并指导他们按照这样的流程去处理后端基础类:

- 在 model.entity 包下新建一个名为 BonusLog 的类,用于对应数据库中的 t_bonus_log 表。这个类将作为后端处理积分日志的实体对象,包含用户 ID、积分变动量、变动时间等关键字段。
- 在 enums 包下新建 BonusActionEnum 枚举,用来列举所有可能的积分变更行为,如

"注册奖励""分享奖励"等。这样的枚举有助于在代码中明确积分变更的原因,增加代码的可读性和可维护性。

- 在 mapper 包下新建 BonusLogMapper 接口,用于与数据库进行交互,实现 BonusLog 实体对象的增、删、改、查操作。Mapper 层是 MyBatis 框架的核心,它将 SQL 语句与 Java 代码分离,提高了开发效率。
- 在 vo 包下新建 BonusLogVO 类,这个类将用于封装返回给前端的积分日志数据。与 BonusLog 实体类不同,BonusLogVO 会根据前端的需求进行定制,只包含前端关心的字段,并且可能会包含一些额外的计算或格式化后的数据。
- 在 convert 包下新建 BonusLogConvert 类,用于实现 BonusLog 和 BonusLogVO 之间的转换。

接下来是 Service 层的业务实现。波哥让小工在 service 包下新建 BonusLogService 接口,并在 service.impl 包下实现这个接口。在 BonusLogServiceImpl 类中,他指导小工实现了查询积分日志的方法。这个方法会调用 BonusLogMapper 提供的数据库查询操作,并将查询结果转换为 BonusLogVO 对象列表返回给前端。为了方便前端进行分类展示,返回的数据结构是根据前端需求定制的。

同时,还要在 UserController 接口中新增一个用户积分详情接口。这个接口会调用 BonusLogService 提供的查询方法,并将查询结果返回给前端。通过这个接口,前端可以获取用户的积分日志详情,并根据需要进行展示。

1.2 任务分析

1.2.1 后端基础类实现

1.2.1.1 PO 实体创建

model.entity 包下新建 BonusLog,对应数据库的 t_bonus_log 表。

```
@Data
@TableName("t_bonus_log")
public class BonusLog {
    @TableId(type = IdType.AUTO)
    private Integer pkId;
    private Integer userId;
    /**
     * @see top.ssy.share.admin.enums.BonusActionEnum
     */
    private String content;
    private Integer bonus;
    @TableField(value = "delete_flag", fill = FieldFill.INSERT)
    @TableLogic
```

```
    private Integer deleteFlag;
    @TableField(value = "update_time", fill = FieldFill.INSERT_UPDATE)
    private LocalDateTime updateTime;
    @TableField(value = "create_time", fill = FieldFill.INSERT)
    private LocalDateTime createTime;
}
```

1.2.1.2 BonusActionEnum 积分情况枚举

enums 包下新建 BonusActionEnum 枚举,列举积分的变更行为。

```
@Getter
public enum BonusActionEnum {

    RESOURCE_AUDIT_PASS(10, "资源审核通过"),

    RESOURCE_EXCHANGE(0, "资源兑换"),

    DAILY_SIGN(5, "每日签到"),

    RESOURCE_BE_EXCHANGED(5, "资源被兑换"),
    ;

    private final Integer bonus;
    private final String desc;

    BonusActionEnum(Integer bonus, String desc) {
        this.bonus = bonus;
        this.desc = desc;
    }
}
```

1.2.1.3 Mapper 层创建

mapper 包下新建 BonusLogMapper 接口。

```
public interface BonusLogMapper extends BaseMapper<BonusLog> {

}
```

1.2.1.4 VO 视图封装

vo 包下新建 BonusLogVO,返回给前端的积分日志视图对象。

```java
@Data
@Schema(name = "BonusLogVO", description = "积分日志返回 vo")
public class BonusLogVO {
    @Schema(name = "pkId", description = "主键")
    private Integer pkId;
    @Schema(name = "userId", description = "用户 id")
    private Integer userId;
    @Schema(name = "bonus", description = "积分")
    private Integer bonus;
    @Schema(name = "content", description = "描述")
    private String content;
    @Schema(name = "createTime", description = "创建时间")
    @JsonFormat(pattern = "yyyy-MM-dd HH:mm:ss", timezone = "GMT+8")
    private LocalDateTime createTime;
}
```

1.2.1.5 Convert 实体转换

convert 包下新建 BonusLogConvert,用于 BonusLogConvert 类转换。

```java
@Mapper
public interface BonusLogConvert {
    BonusLogConvert INSTANCE = Mappers.getMapper(BonusLogConvert.class);

    List<BonusLogVO> convert(List<BonusLog> list);
}
```

1.2.1.6 Service 类创建

1. service 包下新建 BonusLogService 接口。

```java
public interface t_bonus_log extends IService<BonusLog> {



}
```

2. service.impl 包下新建 BonusLogServiceImpl 类,实现 BonusLogService 接口。

```java
@Slf4j
@Service
```

```java
@AllArgsConstructor
public class BonusLogServiceImpl extends ServiceImpl<BonusLogMapper, BonusLog> implements BonusLogService {

}
```

1.2.2 Service层业务实现

1. BonusLogService新增查询方法,这里的返回值是和前端讨论后得出的结构,为了方便前端进行分类展示。

```java
Map<String, List<BonusLogVO>> userBonusResult(Integer userId);
```

2. BonusLogServiceImpl实现方法。

```java
@Override
public Map<String, List<BonusLogVO>> userBonusResult(Integer userId) {
    LambdaQueryWrapper<BonusLog> wrapper = new LambdaQueryWrapper<>();
    wrapper.eq(userId != null, BonusLog::getUserId, userId);
    List<BonusLogVO> voList = BonusLogConvert.INSTANCE.convert(list(wrapper));
    return voList.stream().collect(Collectors.groupingBy(BonusLogVO::getContent));
}
```

1.2.3 Controller层接口实现

UserController接口新增用户积分详情接口。

```java
@GetMapping("bonus/list")
@Operation(summary = "积分列表")
@PreAuthorize("hasAuthority('sys:user:bonus')")
public Result<Map<String, List<BonusLogVO>>> bonusList(@RequestParam Integer userId) {
    return Result.ok(bonusLogService.userBonusResult(userId));
}
```

2 积分管理模块前端实现

2.1 任务描述

完成了积分管理模块后端部分后,波哥开始带领小南和小工转向积分管理模块的前端实现。

首先,波哥指导小南搭建用户积分的静态页面组件。他建议使用 Element Plus 的 Collapse 折叠面板组件来呈现积分列表,因为这样可以提供一个清晰可折叠的视图。小南根据 Element Plus 的官方文档,复制了 Collapse 组件的参考代码,并稍作调整以适应项目风格。调整完成后,她刷新页面查看效果,确保组件的静态展示符合设计要求。

接下来需要将积分组件集成到用户管理页面中。波哥让小工将积分组件导入用户管理页面的 Vue 组件中,并在页面上添加一个按钮,用于触发积分信息的展示。同时还强调了属性绑定和方法配置的重要性,确保按钮点击时能够正确触发积分提示框的显示。

在完成了积分组件的集成后,需要在 src/api/modules/user/index.ts 文件中新增一个用于获取用户积分日志的接口。这个接口需要接收用户的编号作为参数,并返回对应的积分日志数据。

有了接口定义后,就可以实现积分数据的获取和展示。在用户点击积分按钮时,首先调用刚才定义的接口获取用户积分信息。在数据返回之前,页面会弹出一个提示框来显示积分详情。波哥特别强调,这个弹出框需要有一个默认打开的索引,并将其设置为返回数据的第一个键(如果存在的话)。

为了处理返回的数据,波哥指导小工使用计算属性来监听数据的变化。如果数据存在,就遍历并展示积分的详细信息;如果数据不存在,则显示"暂无数据"的提示。在处理数据时,波哥还特别提到了对象结构和数组结构的遍历问题。他告诉小工,由于返回的数据可能是一个嵌套的对象结构,因此需要特殊处理。第一层遍历的是对象的键,而第二层遍历的则是对应键下的数组。如果某个键不存在,就需要显示"暂无数据"的提示。

2.2 任务分析

首先,创建业务组件,路径为:src\views\User\components\UserScoreDialog.vue。

2.2.1 搭建静态效果

搭建用户积分的静态页面组件使用 Element Plus 中 Collapse 折叠面板组件,直接复制组件的参考代码到页面中,查看页面效果。

```
<template>
  <Dialog :model-value="dialogVisible" :title="dialogProps.title" :fullscreen="dialogProps.fullscreen" :max-height="dialogProps.maxHeight" :cancel-dialog="cancelDialog">
    <div :style="'width: calc(100% - ' + dialogProps.labelWidth! / 2 + 'px)'">
      <el-collapse v-model="activeName" @change="handleChange">
        <el-collapse-item title="Consistency" name="1">
          <div> Consistent with real life: in line with the process and logic of real life, and comply with languages and habits that the users are used to; </div>
          <div> Consistent within interface: all elements should be consistent, such as: design style, icons and texts, position of elements, etc </div>
```

```
        </el-collapse-item>
        <el-collapse-item title="Feedback" name="2">
          <div>Operation feedback: enable the users to clearly perceive their operations by style updates and interactive effects;</div>
          <div>Visual feedback: reflect current state by updating or rearranging elements of the page</div>
        </el-collapse-item>
        <el-collapse-item title="Efficiency" name="3">
          <div>Simplify the process: keep operating process simple and intuitive;</div>
          <div>Definite and clear: enunciate your intentions clearly so that the users can quickly understand and make decisions;</div>
          <div>Easy to identify: the interface should be straightforward, which helps the users to identify and frees them from memorizing and recalling</div>
        </el-collapse-item>
        <el-collapse-item title="Controllability" name="4">
          <div>Decision making: giving advices about operations is acceptable, but do not make decisions for the users;</div>
          <div>Controlled consequences: users should be granted the freedom to operate, including canceling, aborting or terminating current operation</div>
        </el-collapse-item>
      </el-collapse>
    </div>
    <template #footer>
      <slot name="footer">
        <el-button @click="cancelDialog">取消</el-button>
      </slot>
    </template>
  </Dialog>
</template>

<script setup lang="ts">
import { Dialog } from '@/components/Dialog'
import { ref } from 'vue'
// 定义弹出框类型
interface DialogProps {
  title: string
  isView: boolean
  fullscreen?: boolean
```

```
  row: any
  labelWidth?: number
  maxHeight?: number | string
  api?: (params: any) => Promise<any>
  getTableList?: () => Promise<any>
}
// 弹出框是否显示
const dialogVisible = ref(false)
// 弹出框属性
const dialogProps = ref<DialogProps>({
  isView: false,
  title: '',
  row: {},
  labelWidth: 160,
  fullscreen: true,
  maxHeight: '500px'
})

// 接收父组件传过来的参数
const acceptParams = async (params: DialogProps) => {
  params.row = { ...dialogProps.value.row, ...params.row }
  dialogProps.value = { ...dialogProps.value, ...params }
  dialogVisible.value = true
}

defineExpose({
  acceptParams
})

// 关闭弹框
const cancelDialog = () => {
  dialogVisible.value = false
  activeName.value = ''
}

// 展开项 高亮
const activeName = ref('1')
```

```
// 点击切换展开项
const handleChange = (val: string) => {
  activeName.value = val
}
</script>

<style scoped lang="less"></style>
```

2.2.2 父组件调用积分组件

导入积分组件到用户管理页面中,绑定积分信息的按钮,配置对应属性及方法,点击按钮弹出用户积分弹框。

```
<template>
  <div class="table-box">
    <!-- 表格操作 -->
    <template #operation="scope">
      <el-button type="primary" link :icon="List" @click="openUserScoreDialog('积分信息', scope.row)" v-hasPermi="['sys:user:bonus']">积分信息</el-button>
    </template>
  </ProTable>
  <UserScoreDialog ref="userScoreDialog" />
  </div>
</template>

<script setup lang="tsx" name="UserManager">
import { List } from '@element-plus/icons-vue'
// 用户积分实例
const userScoreDialog = ref()
// 弹框 传递标题及当前行信息
const openUserScoreDialog = (title: string, row: UserType) => {
  // 合并对应属性
  let params = {
    title,
    row: { ...row },
    isView: title === '积分信息',
    getTableList: proTable.value.getTableList,
    maxHeight: '400px'
  }
```

```
    // 打开弹框
    userScoreDialog.value.acceptParams(params)
}
</script>
```

点击"积分信息"按钮即可查看封装好的用户积分弹框,后续我们需要基于这个弹框进行修改(图1)。

图1 积分信息

2.2.3 定义获取积分接口

在 src/api/modules/user/index.ts 中新增获取用户积分日志的接口,需要传递用户的编号进行积分获取。

```
import http from '@/api'
import { _API } from '@/api/axios/servicePort'

/**
 * @name 用户管理模块
 */
export const UserApi = {
    // 用户积分日志
    getUserPriceLog: (userId: number) => http.get(_API + '/user/bonus/list?userId=' + userId)
}
```

2.2.4 获取用户积分信息

当用户点击按钮后,在页面弹框显示之前,需要要获取相关数据。弹出框需要有默认打开的索引,将其赋值为数据的第一个键。

由于获取的数据是一个对象结构,我们需要判断是否存在对应的数据。因此,可以通过计算属性来监听是否有数据。如果有数据,就展示相关内容;如果没有数据,则显示"暂无数据"。后端返回数据结构如下:

```
"code": 0,
"msg": "success",
"data": {
    "每日签到": [
        {
            "pkId": 54,
            "userId": 19,
            "bonus": 5,
            "content": "每日签到",
            "createTime": "2024-03-16 14:10:07"
        }
    ],
    "资源兑换": [
        {
            "pkId": 55,
            "userId": 19,
            "bonus": -2,
            "content": "资源兑换",
            "createTime": "2024-03-16 14:10:14"
        }
    ]
}
}
```

我们遍历的时候第一层遍历的是一个对象,对"键值对"需要重点处理,如果这个对象有键就显示对应的值;如果没有,就显示"暂无数据"。

```
<template>
  <Dialog :model-value="dialogVisible" :title="dialogProps.title" :fullscreen="dialogProps.fullscreen" :max-height="dialogProps.maxHeight" :cancel-dialog="cancelDialog">
    <div :style="'width: calc(100% - ' + dialogProps.labelWidth! / 2 + 'px)'">
      <el-collapse accordion v-model="activeName" @change="handleChange" v-if="keys.length > 0">
        <el-collapse-item v-for="(item, key) in userScoreInfo" :title="key" :name="key" :key="key">
          <div class="flex items-center justify-between mb-4 border-b py-2" v-for="(child, cindex) in item" :key="cindex">
            <div class="left flex items-center">
              <span>{{ child.content }}</span>
```

```
            <strong class="ml-4 text-md text-[#009688]">{{ child.bonus }}分</strong>
          </div>
          <div class="right">
            <span>{{ child.createTime }}</span>
          </div>
        </div>
      </el-collapse-item>
    </el-collapse>
    <div class="flex items-center justify-center" v-else>
      <el-empty description="暂无数据" />
    </div>
  </div>
 </Dialog>
</template>

<script setup lang="ts">
import { UserApi } from '@/api/modules/user'
import { ref, computed } from 'vue'


const userScoreInfo = ref<any>({})
// 获取对象的 key 判断是否有数据
const keys = computed(() => {
  return Object.keys(userScoreInfo.value)
})


// 接收父组件传过来的参数
const acceptParams = async (params: DialogProps) => {
  params.row = { ...dialogProps.value.row, ...params.row }
  dialogProps.value = { ...dialogProps.value, ...params }
  // 组件刚加载获取数据
  let { data } = (await UserApi.getUserPriceLog(params.row.pkId)) as any
  // 设置数据
  userScoreInfo.value = data || {}
  // 设置默认激活值为第一个键
  activeName.value = Object.keys(data)[0] || ''

  dialogVisible.value = true
```

}

const activeName = ref( )
</script>

点击积分信息按钮后会展示积分信息,展示效果有两种,分别为有数据(图2)和没有数据(图3)。

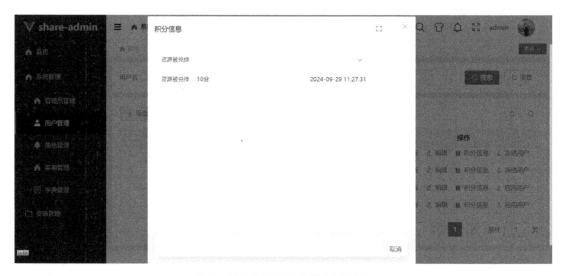

图 2 积分信息展示效果(有数据)

图 3 积分信息展示效果(没有数据)

3 任务总结

在本次积分管理模块的后端实现任务中,我们完成了后端基础类、数据库交互层、服务层以及控制器层的开发,确保了积分管理功能的全面性和用户友好性。

在后端部分,我们构建了坚实的数据模型和业务逻辑,通过实体类、枚举、Mapper 层、服务层和控制器层的协同工作,实现了积分数据的准确处理和高效交互。同时,我们注重代码的规范性和可维护性,为后续的扩展和优化打下了坚实的基础。

在前端部分,我们充分利用了现代前端框架和组件库,快速搭建了用户友好的积分管理界面。我们通过页面搭建、组件调用、接口定义与数据获取、数据展示以及优化与细节处理等环节的工作,成功将后端数据以直观、清晰的方式呈现给用户,并提供了便捷的积分查看和管理功能。

任务六

通知管理模块开发

波哥叫来小南和小工说道:"大家注意一下,我们现在开始新的开发任务——通知管理模块。这个任务包含后端实现和前端实现两个部分,我们需要确保数据的准确性和前后端的顺畅交互。虽然我们预计没有太大的难点,但还是要保持专注和细致,确保每一个功能点都经过充分的测试。"

小南和小工齐声说:"明白,波哥!我们会尽快完成任务的。"

◇ 任务点

- 通知管理模块的后端实现;
- 通知管理模块的前端实现。

◇ 任务计划

- 任务内容:通知管理模块的开发;
- 任务耗时:预计完成时间为 1 h;
- 任务难点:无。

1 通知管理模块后端实现

1.1 任务描述

为了高效完成通知管理模块的后端实现任务,波哥明确了几个关键的步骤和组件。

在 model.dto 包下新建一个 NoticeEditDTO 类。这个类将用于接收前端发送的新增和编辑公告的请求数据,确保数据格式与后端处理一致。

在 model.vo 包下新建一个 NoticeVO 类。这个类将用于封装我们返回给前端的通知列表数据,确保前端能够方便地展示这些数据。

在 convert 包下新建一个 NoticeConvert 接口。这个接口将定义一些方法,用于在 Notice 实体和 NoticeVO、NoticeEditDTO 之间进行数据转换。

在 model.query 包下新建一个 NoticeQuery 类,这个类将继承查询基类,并定义一些用于条件查询公告的属性。

在 mapper 包下创建一个 NoticeMapper 接口,定义分页条件查询等方法。然后,在

resources/mapper 目录下创建一个 NoticeMapper.xml 文件，对刚才定义的方法进行 SQL 实现。

完成 Mapper 层后，在 service 包下创建一个 NoticeService 接口并实现基础接口。

在 controller 层下新建一个 NoticeController 接口。在接口方法中，调用 Service 层的服务来实现具体的业务逻辑。

1.2 任务分析

1.2.1 DTO 实体创建

在 model.dto 包下新建 NoticeEditDTO 类，用于新增和编辑公告。

```
@Data
@Schema(name = "NoticeEditDTO", description = "公告编辑 DTO")
public class NoticeEditDTO {

    @Schema(name = "pkId", description = "主键")
    private Integer pkId;
    @Schema(name = "title", description = "标题")
    private String title;
    @Schema(name = "cover", description = "封面")
    private String cover;
    @Schema(name = "content", description = "内容")
    private String content;
    @Schema(name = "adminId", description = "管理员 ID")
    private Integer adminId;
    @Schema(name = "isTop", description = "是否置顶")
    private Integer isTop;
    @Schema(name = "isSwiper", description = "是否轮播")
    private Integer isSwiper;
    @Schema(name = "deleteFlag", description = "删除标识")
    private Integer deleteFlag;
}
```

1.2.2 VO 返回实体创建

在 model.vo 包下新建 NoticeVO 类，用于返回通知列表。

```
@Data
@Schema(name = "NoticeVO", description = "公告返回 vo")
public class NoticeVO {
```

@Schema(description = "主键")
private Integer pkId;
@Schema(description = "标题")
private String title;
@Schema(description = "封面")
private String cover;
@Schema(description = "内容")
private String content;
@Schema(description = "管理员id")
private Integer adminId;
@Schema(description = "管理员名称")
private String adminName;
@Schema(description = "是否置顶")
private Integer isTop;
@Schema(description = "是否轮播")
private Integer isSwiper;
@Schema(name = "createTime", description = "创建时间")
@JsonFormat(pattern = "yyyy-MM-dd HH:mm:ss", timezone = "GMT+8")
private LocalDateTime createTime;
}

1.2.3 Convert 转换创建

convert 包下新建 NoticeConvert 接口，用作公告相关类的实体转换。

```
@Mapper
public interface NoticeConvert {
    NoticeConvert INSTANCE = Mappers.getMapper(NoticeConvert.class);

    NoticeVO convert(Notice notice);

    Notice convert(NoticeEditDTO dto);
}
```

1.2.4 Query 查询实体创建

model.query 包下新建 NoticeQuery 类，继承查询基类，用于按条件查询公告。

```
package top.ssy.share.admin.model.query;
```

```java
import io.swagger.v3.oas.annotations.media.Schema;
import lombok.Data;
import lombok.EqualsAndHashCode;
import top.ssy.share.admin.common.model.Query;


@EqualsAndHashCode(callSuper = true)
@Data
@Schema(name = "NoticeQuery", description = "公告查询")
public class NoticeQuery extends Query {
    @Schema(name = "标题")
    private String title;
    @Schema(name = "是否置顶")
    private Integer isTop;
    @Schema(name = "是否轮播")
    private Integer isSwiper;
}
```

1.2.5 Mapper 层实现

1. mapper 包下创建 NoticeMapper 接口,定义分页条件查询方法。

```java
package top.ssy.share.admin.mapper;

import com.baomidou.mybatisplus.core.mapper.BaseMapper;
import com.baomidou.mybatisplus.extension.plugins.pagination.Page;
import org.apache.ibatis.annotations.Param;
import top.ssy.share.admin.model.entity.Notice;
import top.ssy.share.admin.model.query.NoticeQuery;
import top.ssy.share.admin.model.vo.NoticeVO;

import java.util.List;

public interface NoticeMapper extends BaseMapper<Notice> {

    List<NoticeVO> getNoticePage(Page<NoticeVO> page, @Param("query") NoticeQuery query);
}
```

2. resources/mapper 下创建 NoticeMapper.xml，对分页方法进行实现。

```xml
<?xml version="1.0" encoding="UTF-8"?>
<!DOCTYPE mapper PUBLIC "-//mybatis.org//DTD Mapper 3.0//EN" "http://mybatis.org/dtd/mybatis-3-mapper.dtd">
<mapper namespace="top.ssy.share.admin.mapper.NoticeMapper">

    <select id="getNoticePage" resultType="top.ssy.share.admin.model.vo.NoticeVO">
        SELECT tn.*, sm.username AS adminName FROM t_notice tn
        LEFT JOIN sys_manager sm ON tn.admin_id = sm.pk_id
        WHERE tn.delete_flag = 0
        <if test="query.title != null and query.title != '' ">
            AND tn.title LIKE concat('%',#{query.title},'%')
        </if>
        <if test="query.isTop != null">
            AND tn.is_top = #{query.isTop}
        </if>
        <if test="query.isSwiper != null">
            AND tn.is_swiper = #{query.isSwiper}
        </if>
        ORDER BY tn.create_time DESC
    </select>

</mapper>
```

1.2.6 Service 层业务实现

1. service 包下创建 NoticeService 接口，实现基础接口。定义三个方法，分别为分页查询、新增和修改公告、批量删除公告。

```java
package top.ssy.share.admin.service;

import com.baomidou.mybatisplus.extension.service.IService;
import top.ssy.share.admin.common.result.PageResult;
import top.ssy.share.admin.model.dto.NoticeEditDTO;
import top.ssy.share.admin.model.entity.Notice;
import top.ssy.share.admin.model.query.NoticeQuery;
import top.ssy.share.admin.model.vo.NoticeVO;

import java.util.List;
```

```java
public interface NoticeService extends IService<Notice> {

    PageResult<NoticeVO> page(NoticeQuery query);

    void saveAndEdit(NoticeEditDTO dto);

    void delete(List<Integer> id);
}
```

2. service.impl 包下新建 NoticeServiceImpl 实现类,实现接口方法。

```java
package top.ssy.share.admin.service.impl;

import com.baomidou.mybatisplus.extension.plugins.pagination.Page;
import com.baomidou.mybatisplus.extension.service.impl.ServiceImpl;
import lombok.AllArgsConstructor;
import org.springframework.stereotype.Service;
import top.ssy.share.admin.common.result.PageResult;
import top.ssy.share.admin.convert.NoticeConvert;
import top.ssy.share.admin.mapper.NoticeMapper;
import top.ssy.share.admin.model.dto.NoticeEditDTO;
import top.ssy.share.admin.model.entity.Notice;
import top.ssy.share.admin.model.query.NoticeQuery;
import top.ssy.share.admin.model.vo.NoticeVO;
import top.ssy.share.admin.service.NoticeService;

import java.util.List;

@Service
@AllArgsConstructor
public class NoticeServiceImpl extends ServiceImpl<NoticeMapper, Notice> implements NoticeService {

    @Override
    public PageResult<NoticeVO> page(NoticeQuery query) {
        Page<NoticeVO> page = new Page<>(query.getPage(), query.getLimit());
        List<NoticeVO> list = baseMapper.getNoticePage(page, query);
        return new PageResult<>(list, page.getTotal());
    }
```

```
@Override
public void saveAndEdit(NoticeEditDTO dto){
    Notice newNotice = NoticeConvert.INSTANCE.convert(dto);
    if(newNotice.getPkId() == null){
        save(newNotice);
    } else {
        updateById(newNotice);
    }
}


@Override
public void delete(List<Integer> id){
    baseMapper.deleteBatchIds(id);
}
}
```

1.2.7　Controller 层接口实现

controller 层下新建 NoticeController 接口。定义查询、新增或修改、删除三个 API 接口，调用 Service 服务。

```
package top.ssy.share.admin.controller;

import io.swagger.v3.oas.annotations.Operation;
import io.swagger.v3.oas.annotations.tags.Tag;
import jakarta.validation.Valid;
import lombok.AllArgsConstructor;
import org.springframework.security.access.prepost.PreAuthorize;
import org.springframework.web.bind.annotation.*;
import top.ssy.share.admin.common.result.PageResult;
import top.ssy.share.admin.common.result.Result;
import top.ssy.share.admin.model.dto.NoticeEditDTO;
import top.ssy.share.admin.model.query.NoticeQuery;
import top.ssy.share.admin.model.vo.NoticeVO;
import top.ssy.share.admin.security.cache.TokenStoreCache;
import top.ssy.share.admin.security.user.ManagerDetail;
import top.ssy.share.admin.service.NoticeService;
```

```java
import java.util.List;

@RestController
@AllArgsConstructor
@Tag(name = "公告管理", description = "公告管理")
@RequestMapping("/notice")
public class NoticeController {
    private final NoticeService noticeService;
    private final TokenStoreCache tokenStoreCache;

    @PostMapping("/page")
    @Operation(summary = "分页")
    @PreAuthorize("hasAuthority('sys:notice:view')")
    public Result<PageResult<NoticeVO>> page(@RequestBody @Valid NoticeQuery query) {
        return Result.ok(noticeService.page(query));
    }


    @PostMapping("saveAndEdit")
    @Operation(summary = "新增或修改")
    @PreAuthorize("hasAnyAuthority('sys:notice:add','sys:notice:edit')")
    public Result<String> saveAndEdit(@RequestBody @Valid NoticeEditDTO dto, @RequestHeader("Authorization") String token) {
        ManagerDetail user = tokenStoreCache.getUser(token);
        dto.setAdminId(user.getPkId());
        noticeService.saveAndEdit(dto);

        return Result.ok();
    }


    @PostMapping("/delete")
    @Operation(summary = "删除")
    @PreAuthorize("hasAuthority('sys:notice:remove')")
    public Result<String> delete(@RequestBody List<Integer> ids) {
        noticeService.delete(ids);
        return Result.ok();
    }
}
```

2 通知管理模块前端实现

2.1 任务描述

在通知管理模块前端开发的阶段,波哥提到了类型声明的重要性,并指导小南和小工进行操作。

波哥说:"小南,首先你需要在 src/api/interface/index.ts 文件中定义与通知管理相关的类型声明。这有助于我们在调用接口和处理数据时保持数据的类型安全。"

波哥转向小工:"小工,你的任务是新增一个接口文件 src/api/modules/notice/index.ts,用于封装与通知管理相关的所有 API 接口。这将使我们能够更方便地调用后端提供的服务。"

"在此之后,小南开始修改 src/views/Resource/Notice.vue 页面,使其能够展示通知列表,并提供相应的操作按钮。"

在开发中,我们经常会遇到需要重复使用的组件,比如弹框。因此,需要在 views/Resource/components 目录下新增几个弹框组件,包括标签弹框、资源弹框、公告弹框和分类弹框。这些组件在后续的开发中可能会多次使用。

对于通知的编辑和查看功能,需要在 src/views/Resource/components/NoticeDialog.vue 弹框组件中添加。用户可以通过这个弹框来查看通知的详细信息,也可以对通知进行编辑。

删除功能包括批量删除和单个删除。我们需要自定义作用域插槽,将需要删除的通知项进行传递,并绑定好对应的事件。当用户点击删除按钮时,调用后端接口进行删除操作,并在删除后重新获取表格数据以刷新页面。

2.2 任务分析

2.2.1 定义类型声明

在 src/api/interface/index.ts 中添加与通知管理相关的类型声明。

```
/** 通知管理 */
export namespace SysNotice {
  export interface ReqGetNoticeParams extends ReqPage {
    name?: string
  }
  /** 通知列表 */
  export interface ResNoticeList {
    pkId: number
    title: string
```

        cover：string

        content：string

        adminId：number

        isTop：number

        isSwiper：number

        deleteFlag：number

        createTime：string

        updateTime：string

    }

    /** 通知编辑 */

    export interface ReqEditNoticeParams {

        pkId?：number

        title：string

        cover：string

        content：string

        adminId：number

        isTop：number

        isSwiper：number

    }

    /** 新增通知 */

    export interface ReqAddNoticeParams {

        pkId?：number

        title：string

        cover：string

        content：string

        adminId：number

        isTop：number

        isSwiper：number

    }

}

2.2.2 添加接口

新增 src/api/modules/notice/index.ts,封装与通知管理相关的接口。

```
import { ResPage, SysNotice } from '@/api/interface/index'
import { _API } from '@/api/axios/servicePort'
import http from '@/api'
```

```typescript
// 通知列表
export const getNoticePage = (params: SysNotice.ReqGetNoticeParams) => {
  return http.post<ResPage<SysNotice.ResNoticeList>>(_API + '/notice/page', params)
}

// 添加通知
export const addNotice = (params: SysNotice.ReqAddNoticeParams) => {
  return http.post(_API + '/notice/saveAndEdit', params)
}

// 编辑通知
export const editNotice = (params: SysNotice.ReqEditNoticeParams) => {
  return http.post(_API + '/notice/saveAndEdit', params)
}

// 删除通知
export const deleteNotice = (params: number[]) => {
  return http.post(_API + '/notice/delete', params)
}
```

2.2.3 通知展示页面

修改 src/views/Resource/Notice.vue,通知管理页面。

```vue
<template>
  <div class="table-box">
    <ProTable ref="proTable" title="通知列表" :columns="columns" :requestApi="getTableList" :initParam="initParam" :dataCallback="dataCallback">
      <!-- 表格 header 按钮 -->
      <template #tableHeader>
        <el-button type="primary" :icon="CirclePlus" @click="openDrawer('新增')" v-hasPermi="['sys:notice:add']">新增通知</el-button>
      </template>
      <!-- 表格操作 -->
      <template #operation="scope">
        <el-button type="primary" link :icon="View" @click="openDrawer('查看', scope.row)" v-hasPermi="['sys:notice:view']">查看</el-button>
        <el-button type="primary" link :icon="EditPen" @click="openDrawer('编辑', scope.row)" v-hasPermi="['sys:notice:edit']">编辑</el-button>
      </template>
```

```
        </ProTable>
        <NoticeDialog ref="dialogRef" />
    </div>
</template>

<script setup name="SysNotice" lang="tsx">
import { ref, reactive } from 'vue'
import { SysNotice } from '@/api/interface'
import { ColumnProps } from '@/components/ProTable/interface'
import ProTable from '@/components/ProTable/index.vue'
import NoticeDialog from '@/views/Resource/components/NoticeDialog.vue'
import { CirclePlus, EditPen, View } from '@element-plus/icons-vue'
import { getNoticePage, addNotice, editNotice } from '@/api/modules/notice'
import { dictConfigList } from '@/api/modules/dict/dictConfig'

// 获取 ProTable 元素,调用其获取刷新数据方法(还能获取当前查询参数,方便导出携带参数)
const proTable = ref()

// 如果表格需要初始化请求参数,可以直接定义参数并传给 ProTable(之后每次请求都会自动带上该
// 参数,此参数更改之后也会一直带上,改变此参数会自动刷新表格数据)
const initParam = reactive({})

// dataCallback 是对于返回的表格数据做处理,如果你后台返回的数据不是 datalist && total 这些字段,
// 那么你可以在这里处理成这些字段
const dataCallback = (data: any) => {
    return {
        list: data.list,
        total: data.total
    }
}

// 如果你想在请求之前对当前请求参数做一些操作,可以自定义如下函数:params 为当前所有的请求参
// 数(包括分页),最后返回请求列表接口
// 默认不做操作就直接在 ProTable 组件上绑定 :requestApi="getUserList"
const getTableList = (params: any) => {
    let newParams = { ...params }
    return getNoticePage(newParams)
}

// 表格配置项
```

```
const columns: ColumnProps[] = [
  { type: 'selection', fixed: 'left', width: 60 },
  {
    prop: 'title',
    label: '通知标题',
    showOverflowTooltip: true,
    search: { el: 'input' }
  },
  {
    prop: 'isTop',
    label: '是否置顶',
    width: 100,
    // enum: [
    //     { title: '是', value: 1 },
    //     { title: '否', value: 0 }
    // ],
    enum: () => dictConfigList('isTop'),
    search: { el: 'select', props: { filterable: true } },
    fieldNames: { label: 'title', value: 'value' },
    render: (scope) => {
      return <el-tag type={scope.row.isTop === 0 ? 'success' : 'warning'}>{scope.row.isTop === 1 ? '是' : '否'}</el-tag>
    }
  },
  {
    prop: 'isSwiper',
    label: '是否轮播',
    width: 100,
    // enum: [
    //     { title: '是', value: 1 },
    //     { title: '否', value: 0 }
    // ],
    enum: () => dictConfigList('isSwiper'),
    search: { el: 'select', props: { filterable: true } },
    fieldNames: { label: 'title', value: 'value' },
    render: (scope) => {
      return (
        <el-tag type={scope.row.isSwiper === 1 ? 'primary' : 'info'} effect>
          {scope.row.isSwiper === 1 ? '是' : '否'}
        </el-tag>
      )
```

```
      }
    },
    {
      prop: 'cover',
      label: '封面',
      width: 100,
      render: (scope) => {
        return (
          <div class={['flex', 'justify-center', 'p-1']}>
            {scope.row.cover ? <el-image style="width: 50px; height 50px" src={scope.row.cover} fit="cover"></el-image> : '暂无图片'}
          </div>
        )
      }
    },
    {
      prop: 'content',
      label: '通知内容'
    },
    {
      prop: 'createTime',
      label: '创建时间',
      width: 200
    },
    { prop: 'operation', label: '操作', fixed: 'right', width: 330 }
]

// 打开 drawer(新增、查看、编辑)
const dialogRef = ref()
const openDrawer = (title: string, row: Partial<SysNotice.ResNoticeList> = {}) => {
  let params = {
    title,
    row: { ...row },
    isView: title === '查看',
    api: title === '新增' ? addNotice : title === '编辑' ? editNotice : '',
    getTableList: proTable.value.getTableList
  }
  dialogRef.value.acceptParams(params)
}
</script>
```

获取后台数据，页面效果如图1所示。

图1 通知管理页面效果图

2.2.4 创建弹框组件

新增组件目录 views/Resource/components，创建 CategoryDialog（分类弹框）、NoticeDialog（公告弹框）、ResourceDialog（资源弹框）、TagDialog（标签弹框）这几个提示框组件（图2），后期需要复用这些弹框组件。

图2 创建提示框组件

2.2.5 新增编辑查看功能

修改通知弹框,编写 src/views/Resource/components/NoticeDialog.vue 弹框组件。

```
<template>
  <Dialog
    :model-value="dialogVisible"
    :title="dialogProps.title"
    :fullscreen="dialogProps.fullscreen"
    :max-height="dialogProps.maxHeight"
    :cancel-dialog="cancelDialog"
    width="900px"
  >
    <div :style="'width: calc(100% - ' + dialogProps.labelWidth! / 2 + 'px)'">
      <el-form
        ref="ruleFormRef"
        label-position="right"
        :label-width="dialogProps.labelWidth + 'px'"
        :rules="rules"
        :model="dialogProps.row"
        :disabled="dialogProps.isView"
        :hide-required-asterisk="dialogProps.isView"
      >
        <el-form-item label="通知标题" prop="title">
          <el-input v-model="dialogProps.row!.title" placeholder="请输入通知标题" clearable></el-input>
        </el-form-item>

        <el-form-item label="是否置顶" prop="isTop">
          <el-radio-group v-model="dialogProps.row!.isTop">
            <el-radio :label="0" border>否</el-radio>
            <el-radio :label="1" border>是</el-radio>
          </el-radio-group>
        </el-form-item>
        <el-form-item label="是否轮播" prop="isSwiper">
          <el-radio-group v-model="dialogProps.row!.isSwiper">
            <el-radio :label="0" border>否</el-radio>
            <el-radio :label="1" border>是</el-radio>
          </el-radio-group>
```

```
        </el-form-item>
        <el-form-item label="轮播图" prop="cover" v-if="dialogProps.row!.isSwiper===1">
          <UploadImg v-model:image-url="dialogProps.row!.cover" width="135px" height="135px" :file-size="5">
            <template #empty>
              <el-icon><Upload /></el-icon>
              <span>请上传轮播图</span>
            </template>
            <template #tip>轮播图大小不能超过 5MB </template>
          </UploadImg>
        </el-form-item>
        <el-form-item label="内容" prop="content">
          <WangEditor v-model:value="dialogProps.row!.content" height="200px" />
        </el-form-item>
      </el-form>
    </div>
    <template #footer>
      <slot name="footer">
        <el-button @click="cancelDialog">取消</el-button>
        <el-button type="primary" v-show="!dialogProps.isView" @click="handleSubmit">确定</el-button>
      </slot>
    </template>
  </Dialog>
</template>

<script setup lang="ts">
import { ref, reactive } from 'vue'
import { ElMessage, FormInstance } from 'element-plus'
import Dialog from '@/components/Dialog'
import UploadImg from '@/components/Upload/Img.vue'
import WangEditor from '@/components/WangEditor/index.vue'
interface DialogProps {
  title: string
  isView: boolean
  fullscreen?: boolean
  row: any
  labelWidth?: number
```

```
  maxHeight?: number | string
  api?: (params: any) => Promise<any>
  getTableList?: () => Promise<any>
}
const dialogVisible = ref(false)
const dialogProps = ref<DialogProps>({
  isView: false,
  title: '',
  row: {},
  labelWidth: 160,
  fullscreen: true,
  maxHeight: '500px'
})

// 接收父组件传过来的参数
const acceptParams = (params: DialogProps): void => {
  if (params.row.isTop == null) {
    params.row.isTop = 0
  }
  if (params.row.isSwiper == null) {
    params.row.isSwiper = 0
  }
  params.row = { ...dialogProps.value.row, ...params.row }
  dialogProps.value = { ...dialogProps.value, ...params }
  dialogVisible.value = true
}

defineExpose({
  acceptParams
})

const rules = reactive({
  title: [
    { required: true, message: '请输入通知标题', trigger: 'blur' },
    {
      min: 2,
      max: 10,
      message: '长度在 2 到 10 个字符',
      trigger: 'blur'
    }
```

```js
    ],
    content: [
      { required: true, message: '请输入通知内容', trigger: 'blur' },
      {
        min: 1,
        max: 200,
        message: '长度在 1 到 200 个字符',
        trigger: 'blur'
      }
    ],
    isTop: [{ required: true, message: '请选择是否置顶', trigger: 'blur' }],
    isSwiper: [{ required: true, message: '请选择是否轮播', trigger: 'blur' }]
})
const ruleFormRef = ref<FormInstance>()
const handleSubmit = () => {
  ruleFormRef.value!.validate(async (valid) => {
    if (!valid) return
    try {
      await dialogProps.value.api!(dialogProps.value.row)
      ElMessage.success({ message: `${dialogProps.value.title}成功!` })
      dialogProps.value.getTableList!()
      dialogVisible.value = false
      ruleFormRef.value!.resetFields()
      cancelDialog(true)
    } catch (error) {
      console.log(error)
    }
  })
}
const cancelDialog = (isClean?: boolean) => {
  dialogVisible.value = false
  let condition = ['查看', '编辑']
  if (condition.includes(dialogProps.value.title) || isClean) {
    dialogProps.value.row = {}
    ruleFormRef.value!.resetFields()
  }
}
</script>

<style scoped lang="less"></style>
```

点击"新增",页面效果如图3所示。

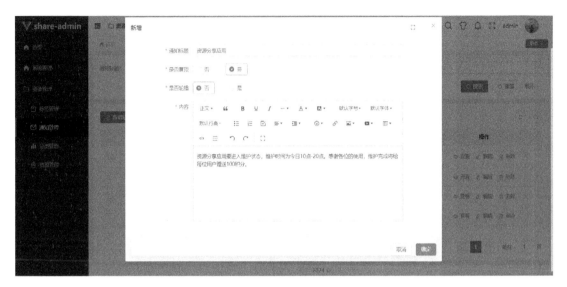

图3 "新增"功能页面效果图

点击"编辑",页面效果如图4所示。

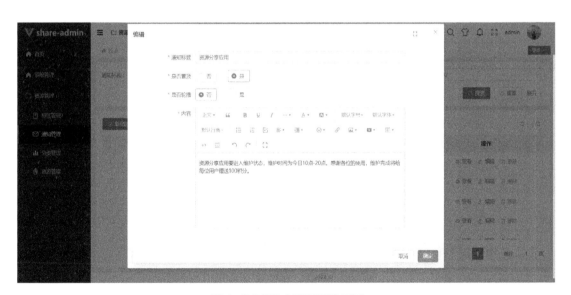

图4 "编辑"功能页面效果图

点击"查看",页面效果如图5所示。

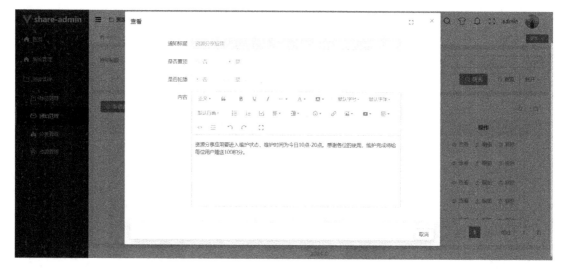

图5 "查看"功能页面效果图

2.2.6 删除功能

删除功能主要包括批量删除与单个删除。我们需要自定义作用域插槽,将需要删除项进行传递,绑定好相应的事件,调用后端接口方法进行删除。删除后,再调用获取表格数据的方法进行页面刷新。

```
<template>
  <div class="table-box">
    <ProTable ref="proTable" title="通知列表" :columns="columns" :requestApi="getTableList" :initParam="initParam" :dataCallback="dataCallback">
      <!-- 表格 header 按钮 -->
      <template #tableHeader>
        <el-button type="primary" :icon="CirclePlus" @click="openDrawer('新增')" v-hasPermi="['sys:notice:add']">新增通知</el-button>
      </template>
      <!-- 表格操作 -->
      <template #operation="scope">
        <el-button type="primary" link :icon="View" @click="openDrawer('查看', scope.row)" v-hasPermi="['sys:notice:view']">查看</el-button>
        <el-button type="primary" link :icon="EditPen" @click="openDrawer('编辑', scope.row)" v-hasPermi="['sys:notice:edit']">编辑</el-button>
        <el-button type="danger" link :icon="Delete" @click="deleteRow(scope.row)" v-hasPermi="['sys:notice:remove']">删除</el-button>
```

```
        </template>
      </ProTable>
      <NoticeDialog ref="dialogRef" />
    </div>
</template>

<script setup name="SysNotice" lang="tsx">
import { Delete } from '@element-plus/icons-vue'
import { useHandleData } from '@/hooks/useHandleData'
import { deleteNotice } from '@/api/modules/notice'

// 删除单个
const deleteRow = async (params: SysNotice.ResNoticeList) => {
  await useHandleData(deleteNotice, [params.pkId], '删除【${params.title}】通知')
  proTable.value.getTableList()
}
</script>
```

点击"删除",页面效果如图6所示。

图6 "删除"确认图

3 任务总结

本次任务成功完成了后台管理系统中通知管理模块的开发,涵盖了从后端数据模型构建到前端交互设计的全过程。

在后端实现上，我们关注数据模型的标准化、数据转换的简化、查询与映射的灵活性、业务逻辑的核心实现以及接口定义的清晰性。通过定义数据传输对象和视图对象，我们确保了前后端数据交互的一致性和效率。同时，通过构建查询实体和 Mapper 层，我们为前端提供了灵活的条件查询和分页功能。在 Service 层，我们实现了通知的分页查询、新增、修改和批量删除等核心业务逻辑，为前端提供了丰富的功能支持。最后，在 Controller 层定义了 API 接口，为前端提供了清晰的调用路径。

在前端实现上，我们注重接口的封装、页面的展示、组件的复用、交互逻辑的实现以及数据的实时更新。通过封装与后端通信的 API 接口，我们简化了前端的调用过程。同时，设计了清晰明了的通知管理页面，包括通知列表展示和基本操作按钮，提升了用户的使用体验。在组件复用方面，我们创建了多个弹框组件，提高了开发效率。在交互逻辑上，我们实现了通知的编辑、查看、删除等功能，进一步提升了用户体验。最后，我们通过 Vue 框架的数据双向绑定机制，确保了页面元素与数据的实时同步更新。

任务七

标签管理模块开发

波哥:"大家好,我们接下来需要完成后台管理系统中标签管理模块的开发,主要包括标签的增、删、改、查功能。这个任务由小南和小工你们两人协作完成。考虑到这是一个相对独立的模块,我们并没有设置特别大的难点。如果遇到任何问题,随时向我反馈。那就开始吧,确保按时完成。"

任务点

- 标签管理模块的后端实现;
- 标签管理模块的前端实现。

任务计划

- 任务内容:标签管理模块的开发;
- 任务耗时:预计完成时间为 1 h;
- 任务难点:无。

1 标签管理模块后端实现

1.1 任务描述

波哥正在为标签管理模块的后端部分拆解开发任务:

- 在 model/entity 子包中新建一个 Tag 实体类,这个类要对应数据库中的 t_tag 表,用于存储标签的详细信息。
- 在 enums 包下创建一个 CommonStatusEnum 枚举类,用于标识标签(以及其他可能使用到的实体)的基础状态,比如激活、禁用等。
- 在 model/dto 子包中新建一个 TagEditDTO 类,这个类将用于前端在新增或修改标签时发送数据。同时,在 model/vo 子包中创建一个 TagVO 类,用于后端将标签数据以友好的方式返回给前端。
- 在 convert 子包中新建一个 TagConvert 接口,并实现这个接口,用于 Tag、TagEditDTO 和 TagVO 之间的数据转换。
- 在 model/query 子包中创建一个 TagQuery 类,这个类需要继承通用的 Query 类,并定义一些用于查询的属性。
- 在 Mapper 层,需要在 mapper 包中创建一个 TagMapper 接口,并定义分页条件查询等方法。然后在 resources/mapper 目录下创建 TagMapper.xml 文件,实现这些接口方法。
- 在 Service 层,需要在 service 包中创建 TagService 接口,并定义增、删、改、查等方法。然后在 service.impl 包中创建 TagServiceImpl 类,实现这些接口方法。
- 在 Controller 层,需要在 controller 包中新建 TagController 接口,并定义与标签管理相关的 API 接口。在接口中调用 Service 层的方法来实现具体的业务逻辑。

小南和小工点头表示明白,并一起开始了标签管理模块后端部分的开发工作。

1.2 任务分析

1.2.1 PO 实体创建

在 model/entity 子包中新建 Tag 实体类,对应 t_tag 表。

```java
@Getter
@Setter
@ToString
@TableName("t_tag")
public class Tag {
    @TableId(type = IdType.AUTO, value = "pk_id")
    private Integer pkId;
    private String title;
    private String description;
    /**
     * @see top.ssy.share.admin.enums.CommonStatusEnum
     */
    private Integer isHot;
    @TableField(value = "delete_flag", fill = FieldFill.INSERT)
    @TableLogic
    private Integer deleteFlag;
    @TableField(value = "update_time", fill = FieldFill.INSERT_UPDATE)
    private LocalDateTime updateTime;
    @TableField(value = "create_time", fill = FieldFill.INSERT)
    private LocalDateTime createTime;
}
```

1.2.2 CommonStatusEnum 枚举创建

在 enums 包下创建 CommonStatusEnum 基础状态枚举类。

```java
@Getter
public enum CommonStatusEnum {

    NO(0,"否"),
    YES(1,"是"),
    ;

    private final Integer status;
    private final String desc;

    CommonStatusEnum(Integer status, String desc) {
        this.status = status;
        this.desc = desc;
    }
}
```

1.2.3 DTO 实体创建

在 model/dto 子包中新建 TagEditDTO 类,用于新增和修改标签。

```java
@Data
@Schema(description = "标签新增修改 dto")
public class TagEditDTO {
    @Schema(name = "pkId", description = "主键")
    private Integer pkId;
    @Schema(name = "title", description = "标题")
    private String title;
    @Schema(name = "description", description = "描述")
    private String description;
    @Schema(name = "isHot", description = "是否热门")
    private Integer isHot;
    @Schema(name = "deleteFlag", description = "删除标识")
    private Integer deleteFlag;
}
```

1.2.4 VO 返回实体创建

在 model/vo 子包中新建 TagVO 类,用于返回给前端标签视图类。

```java
@Data
@Schema(description = "标签返回vo")
public class TagVO {
    @Schema(name = "pkId", description = "主键")
    private Integer pkId;
    @Schema(name = "title", description = "标题")
    private String title;
    @Schema(name = "description", description = "描述")
    private String description;
    @Schema(name = "isHot", description = "是否热门")
    private Integer isHot;
    @Schema(name = "createTime", description = "创建时间")
    @JsonFormat(pattern = "yyyy-MM-dd HH:mm:ss", timezone = "GMT+8")
    private LocalDateTime createTime;
}
```

1.2.5 Convert 转换创建

在 convert 子包中新建 TagConvert 转换接口,用于标签相关的实体类转换。

```java
package top.ssy.share.admin.convert;

import org.mapstruct.Mapper;
import org.mapstruct.factory.Mappers;
import top.ssy.share.admin.model.dto.TagEditDTO;
import top.ssy.share.admin.model.entity.Tag;
@Mapper
public interface TagConvert {

    TagConvert INSTANCE = Mappers.getMapper(TagConvert.class);


    Tag convert(TagEditDTO dto);

}
```

1.2.6 Query 查询实体创建

在 model/query 子包中创建 TagQuery 类,继承 Query 类,用于条件查询。

```
package top.ssy.share.admin.model.query;

import io.swagger.v3.oas.annotations.media.Schema;
import lombok.Data;
import lombok.EqualsAndHashCode;
import top.ssy.share.admin.common.model.Query;


@Data
@Schema(description = "标签查询")
public class TagQuery extends Query {
    @Schema(description = "标题")
    private String title;
    @Schema(description = "是否热门")
    private Integer isHot;
}
```

1.2.7 Mapper 层实现

1. 在 mapper 包创建 TagMapper 接口,定义分页条件查询方法。

```
package top.ssy.share.admin.mapper;

import com.baomidou.mybatisplus.core.mapper.BaseMapper;
import com.baomidou.mybatisplus.extension.plugins.pagination.Page;
import org.apache.ibatis.annotations.Param;
import top.ssy.share.admin.model.entity.Tag;
import top.ssy.share.admin.model.query.TagQuery;
import top.ssy.share.admin.model.vo.TagVO;

import java.util.List;


public interface TagMapper extends BaseMapper<Tag> {
    List<TagVO> getTagPage(Page<TagVO> page, @Param("query") TagQuery query);
}
```

2. 在 resources/mapper 下创建 TagMapper.xml,实现分页方法。

```xml
<?xml version="1.0" encoding="UTF-8"?>
<!DOCTYPE mapper PUBLIC "-//mybatis.org//DTD Mapper 3.0//EN" "http://mybatis.org/dtd/mybatis-3-mapper.dtd">
<mapper namespace="top.ssy.share.admin.mapper.TagMapper">

    <select id="getTagPage" resultType="top.ssy.share.admin.model.vo.TagVO">
        SELECT tt.* FROM t_tag tt
        WHERE tt.delete_flag = 0
        <if test="query.title != null and query.title != '' ">
            AND tt.title LIKE concat('%',#{query.title},'%')
        </if>
        <if test="query.isHot != null ">
            AND tt.is_hot = #{query.isHot}
        </if>
        ORDER BY tt.create_time DESC
    </select>

</mapper>
```

1.2.8 Service 层业务实现

1. 在 service 包中 TagService 接口,定义增、删、改、查方法。

```java
package top.ssy.share.admin.service;

import com.baomidou.mybatisplus.extension.service.IService;
import top.ssy.share.admin.common.result.PageResult;
import top.ssy.share.admin.model.dto.TagEditDTO;
import top.ssy.share.admin.model.entity.Tag;
import top.ssy.share.admin.model.query.TagQuery;
import top.ssy.share.admin.model.vo.TagVO;

import java.util.List;

public interface TagService extends IService<Tag> {

    PageResult<TagVO> page(TagQuery query);
```

```
    void saveAndEdit(TagEditDTO dto);

    void delete(List<Integer> ids);
}
```

2. 在 service.impl 包中新建 TagServiceImpl 类,实现接口方法。

```
package top.ssy.share.admin.service.impl;

import com.baomidou.mybatisplus.extension.plugins.pagination.Page;
import com.baomidou.mybatisplus.extension.service.impl.ServiceImpl;
import org.springframework.stereotype.Service;
import top.ssy.share.admin.common.result.PageResult;
import top.ssy.share.admin.convert.TagConvert;
import top.ssy.share.admin.mapper.TagMapper;
import top.ssy.share.admin.model.dto.TagEditDTO;
import top.ssy.share.admin.model.entity.Tag;
import top.ssy.share.admin.model.query.TagQuery;
import top.ssy.share.admin.model.vo.TagVO;
import top.ssy.share.admin.service.TagService;

import java.util.List;

@Service
public class TagServiceImpl extends ServiceImpl<TagMapper, Tag> implements TagService {
    @Override
    public PageResult<TagVO> page(TagQuery query) {
        Page<TagVO> page = new Page<>(query.getPage(), query.getLimit());
        List<TagVO> list = baseMapper.getTagPage(page, query);
        return new PageResult<>(list, page.getTotal());
    }

    @Override
    public void saveAndEdit(TagEditDTO dto) {
        Tag tag = TagConvert.INSTANCE.convert(dto);
        if (tag.getPkId() == null) {
            save(tag);
        } else {
            updateById(tag);
```

```
        }
    }

    @Override
    public void delete(List<Integer> ids) {
        baseMapper.deleteBatchIds(ids);
    }
}
```

1.2.9 Controller 层接口实现

在 controller 包下新建 TagController 接口,定义 API 接口。

```
package top.ssy.share.admin.controller;

import io.swagger.v3.oas.annotations.Operation;
import io.swagger.v3.oas.annotations.tags.Tag;
import jakarta.validation.Valid;
import lombok.AllArgsConstructor;
import org.springframework.security.access.prepost.PreAuthorize;
import org.springframework.web.bind.annotation.PostMapping;
import org.springframework.web.bind.annotation.RequestBody;
import org.springframework.web.bind.annotation.RequestMapping;
import org.springframework.web.bind.annotation.RestController;
import top.ssy.share.admin.common.result.PageResult;
import top.ssy.share.admin.common.result.Result;
import top.ssy.share.admin.model.dto.TagEditDTO;
import top.ssy.share.admin.model.query.TagQuery;
import top.ssy.share.admin.model.vo.TagVO;
import top.ssy.share.admin.service.TagService;

import java.util.List;

@RestController
@AllArgsConstructor
@Tag(name = "标签管理", description = "标签管理")
@RequestMapping("/tag")
public class TagController {
    private final TagService tagService;
```

```java
@PostMapping("/page")
@Operation(summary = "分页")
@PreAuthorize("hasAuthority('sys:tag:view')")
public Result<PageResult<TagVO>> page(@RequestBody @Valid TagQuery query) {
    return Result.ok(tagService.page(query));
}

@PostMapping("saveAndEdit")
@Operation(summary = "新增或修改")
@PreAuthorize("hasAnyAuthority('sys:tag:add', 'sys:tag:edit')")
public Result<String> saveAndEdit(@RequestBody @Valid TagEditDTO dto) {
    tagService.saveAndEdit(dto);
    return Result.ok();
}

@PostMapping("/delete")
@Operation(summary = "删除")
@PreAuthorize("hasAuthority('sys:tag:remove')")
public Result<String> delete(@RequestBody List<Integer> ids) {
    tagService.delete(ids);
    return Result.ok();
}
}
```

至此,标签管理的后端编写完毕。

2 标签管理模块前端实现

2.1 任务描述

波哥叫来了小南和小工,讨论并分配了标签管理模块前端实现的任务。波哥首先对小工说:"小工,你需要在 src/api/interface/index.ts 中定义与标签管理相关的类型声明,确保我们与后端的数据交互有明确的类型约束。接下来,在 src/api/modules/tag/index.ts 中封装标签管理相关的接口。这些接口将作为前端与后端通信的桥梁,要根据后端提供的 API 进行封装。"

波哥转向小南说:"小南,标签的展示页面已经在 src/views/Resource/Tag.vue 中有了基本的结构。但我们需要对其进行修改,以满足标签管理的需求。你可以根据参考官网文档中的表格参数来修改和完善标签相关的代码。在标签的展示页面中,我们需要实现新增、编辑和查看标签的功能。你可以通过传递的属性来判断页面是哪种操作模式。在打开对话框时，根据操作模式的不同，对表单进行相应的处理。"

波哥又提到了删除功能："删除功能包括批量删除和单个删除。我们要在页面上添加自定义的作用域插槽，将需要删除的项进行传递，并绑定好对应的事件。当点击删除按钮时，调用后端接口方法进行删除操作，并在删除后刷新页面数据。"

小南和小工分工合作，开始了标签管理模块前端部分的开发工作。

2.2 任务分析

2.2.1 定义类型声明

在 src/api/interface/index.ts 中，添加与标签管理相关的类型声明。

```
/** 标签管理 */
export namespace SysTag {
  export interface ReqGetTagParams extends ReqPage {
    name?: string
  }
  /** 标签列表 */
  export interface ResTagList {
    pkId: number
    title: string
    description: string
    isHot: string
    deleteFlag: number
    createTime: string
    updateTime: string
  }
  /** 标签修改 */
  export interface ReqEditTagParams {
    title: string
    description: string
    isHot: string
  }
  /** 新增标签 */
  export interface ReqAddTagParams {
    title: string
    description: string
    isHot: string
  }
}
```

2.2.2 添加接口

在 src/api/modules/tag/index.ts 中新增封装标签管理相关接口。

```ts
import { ResPage, SysTag } from '@/api/interface/index'
import { _API } from '@/api/axios/servicePort'
import http from '@/api'

// 标签列表
export const getTagPage = (params: SysTag.ReqGetTagParams) => {
  return http.post<ResPage<SysTag.ResTagList>>(_API + '/tag/page', params)
}


// 添加标签
export const addTag = (params: SysTag.ReqAddTagParams) => {
  return http.post(_API + '/tag/saveAndEdit', params)
}


// 编辑标签
export const editTag = (params: SysTag.ReqEditTagParams) => {
  return http.post(_API + '/tag/saveAndEdit', params)
}


// 删除标签
export const deleteTag = (params: number[]) => {
  return http.post(_API + '/tag/delete', params)
}
```

2.2.3 标签展示页面

修改 src/views/Resource/Tag.vue 标签组件,可以根据自己的需求进行自定义,里面的具体参数可以查看上一章节,或者参考官网所有参数的含义和效果,修改标签相关代码。

```vue
<template>
  <div class="table-box">
    <ProTable ref="proTable" title="标签列表" :columns="columns" :requestApi="getTableList" :initParam="initParam" :dataCallback="dataCallback">
```

```
        <!-- 表格 header 按钮 -->
        <template #tableHeader>
            <el-button type="primary" :icon="CirclePlus" @click="openDrawer('新增')" v-hasPermi=
"['sys:tag:add']">新增标签</el-button>
        </template>
        <!-- 表格操作 -->
        <template #operation="scope">
            <el-button type="primary" link :icon="View" @click="openDrawer('查看', scope.row)" v-
hasPermi="['sys:tag:view']">查看</el-button>
            <el-button type="primary" link :icon="EditPen" @click="openDrawer('编辑', scope.row)" v-
hasPermi="['sys:tag:edit']">编辑</el-button>
        </template>
    </ProTable>
    <TagDialog ref="dialogRef" />
  </div>
</template>

<script setup lang="tsx" name="SysTag">
import { ref, reactive } from 'vue'
import { SysTag } from '@/api/interface'
import { ColumnProps } from '@/components/ProTable/interface'
import ProTable from '@/components/ProTable/index.vue'
import TagDialog from '@/views/Resource/components/TagDialog.vue'
import { CirclePlus, EditPen, View } from '@element-plus/icons-vue'
import { getTagPage, addTag, editTag } from '@/api/modules/tag'
import { dictConfigList } from '@/api/modules/dict/dictConfig'

// 获取 ProTable 元素,调用其获取刷新数据方法(还能获取当前查询参数,方便导出携带参数)
const proTable = ref()

// 如果表格需要初始化请求参数,直接定义传给 ProTable(之后每次请求都会自动带上该参数,此参数
// 更改之后也会一直带上,改变此参数会自动刷新表格数据)
const initParam = reactive({})

// dataCallback 是对于返回的表格数据做处理,如果你后台返回的数据不是 datalist && total 这些字段,
// 那么你可以在这里处理成这些字段
const dataCallback = (data: any) => {
```

```
  return {
    list: data.list,
    total: data.total
  }
}

// 如果你想在请求之前对当前请求参数做一些操作,可以自定义如下函数:params 为当前所有的请求参
// 数(包括分页),最后返回请求列表接口
// 默认不做操作就直接在 ProTable 组件上绑定 :requestApi = " getUserList"
const getTableList = (params: any) => {
  let newParams = { ...params }
  return getTagPage(newParams)
}
// 表格配置项
const columns: ColumnProps[] = [
  { type: 'selection', fixed: 'left', width: 60 },
  {
    prop: 'title',
    label: '标签名称',
    search: { el: 'input' }
  },
  {
    prop: 'isHot',
    label: '是否热门',
    width: 100,
    // enum: [
    //   { title: '否', value: 0 },
    //   { title: '是', value: 1 }
    // ],
    enum: () => dictConfigList('isHot'),
    search: {
      el: 'select',
      props: { filterable: true }
    },
    fieldNames: { label: 'title', value: 'value' },
    render: (scope) => {
```

```
            return (
                <el-tag type={scope.row.isHot === 1 ? 'success' : 'info'} effect>
                    {scope.row.isHot === 1 ? '是' : '否'}
                </el-tag>
            )
        }
    },
    {
        prop: 'description',
        label: '标签描述'
    },

    {
        prop: 'createTime',
        label: '创建时间',
        width: 200
    },
    { prop: 'operation', label: '操作', fixed: 'right', width: 330 }
]

// 打开 drawer(新增、查看、编辑)
const dialogRef = ref()
const openDrawer = (title: string, row: Partial<SysTag.ResTagList> = {}) => {
    let params = {
        title,
        row: { ...row },
        isView: title === '查看',
        api: title === '新增' ? addTag : title === '编辑' ? editTag : '',
        getTableList: proTable.value.getTableList,
        maxHeight: '200px'
    }
    dialogRef.value.acceptParams(params)
}
</script>
```

获取页面数据后,页面效果如图 1 所示。

图1　标签展示页面效果图

2.2.4　新增、编辑、查看功能

我们在打开对话框的时候,可以通过传递的属性来判断页面是新增、编辑还是查看。如果是新增,需要清空表单;如果是编辑,需要将传递进来的数据进行赋值;如果是查看,需要将传递进来的数据进行赋值并禁用表单。

修改标签弹框组件 src/views/Resource/components/TagDialog.vue。

```
<template>
  <Dialog :model-value="dialogVisible" :title="dialogProps.title" :fullscreen="dialogProps.fullscreen" :max-height="dialogProps.maxHeight" :cancel-dialog="cancelDialog">
    <div :style="'width: calc(100% - ' + dialogProps.labelWidth / 2 + 'px)'">
      <el-form
        ref="ruleFormRef"
        label-position="right"
        :label-width="dialogProps.labelWidth + 'px'"
        :rules="rules"
        :model="dialogProps.row"
        :disabled="dialogProps.isView"
        :hide-required-asterisk="dialogProps.isView"
      >
        <el-form-item label="标签名称" prop="title">
          <el-input v-model="dialogProps.row.title" placeholder="请输入标签名称" maxlength="200"></el-input>
```

```html
            </el-form-item>
            <el-form-item label="标签描述" prop="description">
              <el-input v-model="dialogProps.row.description" :rows="3" type="textarea" placeholder="请输入标签描述" maxlength="200" />
            </el-form-item>
            <el-form-item label="是否热门" prop="isHot">
              <el-radio-group v-model="dialogProps.row.isHot">
                <el-radio :label="0" border>否</el-radio>
                <el-radio :label="1" border>是</el-radio>
              </el-radio-group>
            </el-form-item>
          </el-form>
        </div>
        <template #footer>
          <slot name="footer">
            <el-button @click="cancelDialog">取消</el-button>
            <el-button type="primary" v-show="!dialogProps.isView" @click="handleSubmit">确定</el-button>
          </slot>
        </template>
      </Dialog>
</template>

<script setup lang="ts">
import { ref, reactive } from 'vue'
import { ElMessage, FormInstance } from 'element-plus'
import Dialog from '@/components/Dialog'

interface DialogProps {
  title: string
  isView: boolean
  fullscreen?: boolean
  row: any
  labelWidth?: number
  maxHeight?: number | string
  api?: (params: any) => Promise<any>
  getTableList?: () => Promise<any>
```

```
    roleList?: any
}
const dialogVisible = ref(false)
const dialogProps = ref<DialogProps>({
    isView: false,
    title: '',
    row: {},
    labelWidth: 160,
    fullscreen: true,
    maxHeight: '50vh'
})

// 接收父组件传过来的参数
const acceptParams = (params: DialogProps): void => {
    if (params.row.isHot == null) {
        params.row.isHot = 0
    }
    params.row = { ...dialogProps.value.row, ...params.row }
    dialogProps.value = { ...dialogProps.value, ...params }
    dialogVisible.value = true
}

defineExpose({
    acceptParams
})

const rules = reactive({
    title: [{ required: true, message: '请输入标签名称', trigger: 'blur' }],
    isHot: [{ required: true, message: '请选择是否热门', trigger: 'blur' }],
    description: [
        { required: true, message: '请输入标签描述', trigger: 'blur' },
        { required: true, min: 1, max: 200, message: '请输入标签描述1-200字以内', trigger: 'blur' }
    ]
})
const ruleFormRef = ref<FormInstance>()
const handleSubmit = () => {
```

```
ruleFormRef.value?.validate(async (valid) => {
    if (!valid) return
    try {
        await dialogProps.value.api!(dialogProps.value.row)
        ElMessage.success({ message: `${dialogProps.value.title}成功！` })
        dialogProps.value.getTableList!()
        dialogVisible.value = false
        ruleFormRef.value?.resetFields()
        cancelDialog()
    } catch (error) {
        console.log(error)
    }
})
}
const cancelDialog = () => {
    dialogVisible.value = false
    dialogProps.value.row = {}
    ruleFormRef.value?.resetFields()
}
</script>
```

点击"新增"，页面效果如图 2 所示。

图 2 "新增"页面效果图

点击"编辑",页面效果如图 3 所示。

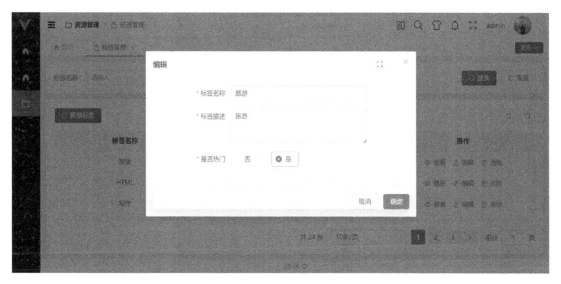

图 3 "编辑"页面效果图

点击"查看",页面效果如图 4 所示。

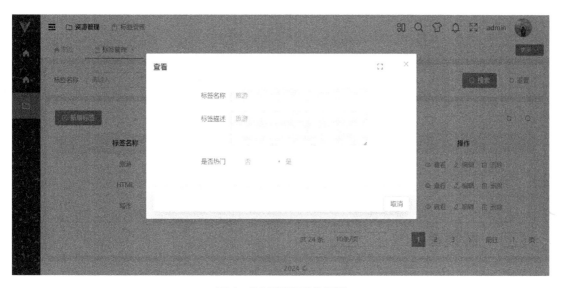

图 4 "查看"页面效果图

2.2.5 删除功能

删除功能主要包括批量删除与单个删除。我们需要自定义作用域插槽,将需要删除项进行传递,绑定好对应的事件,调用后端接口方法进行删除,删除后,再调用获取表格数据的方法进行页面刷新,修改 Tag.vue 页面代码。

```vue
<template>
  <div class="table-box">
    <ProTable ref="proTable" title="标签列表" :columns="columns" :requestApi="getTableList" :initParam="initParam" :dataCallback="dataCallback">
      <!-- 表格 header 按钮 -->
      <template #tableHeader="scope">
        <el-button type="primary" :icon="CirclePlus" @click="openDrawer('新增')" v-hasPermi="['sys:tag:add']">新增标签</el-button>
        <el-button type="danger" :icon="Delete" plain :disabled="!scope.isSelected" @click="deleteBatch(scope.selectedListIds as unknown as number[])" v-hasPermi="['sys:tag:remove']"
        >批量删除标签</el-button>
        >
      </template>
      <!-- 表格操作 -->
      <template #operation="scope">
        <el-button type="danger" link :icon="Delete" @click="deleteRow(scope.row)" v-hasPermi="['sys:tag:remove']">删除</el-button>
      </template>
    </ProTable>
    <TagDialog ref="dialogRef" />
  </div>
</template>
<script setup lang="tsx" name="SysTag">
import { Delete } from '@element-plus/icons-vue'
import { ElMessageBox, ElMessage } from 'element-plus'
import { deleteTag } from '@/api/modules/tag'
import { useHandleData } from '@/hooks/useHandleData'

// 删除单个
const deleteRow = async (params: SysTag.ResTagList) => {
  await useHandleData(deleteTag, [params.pkId], `删除【${params.title}】标签`)
  proTable.value.getTableList()
}

// 批量删除
const deleteBatch = async (ids: number[]) => {
  if (ids.length === 0) {
```

```
    ElMessageBox.alert('请选择需要删除的标签','提示',{type:'warning'})
    return
  }
  ElMessageBox.confirm('是否确认删除选中的标签？','提示',{
    confirmButtonText：'确定',
    cancelButtonText：'取消',
    type：'warning'
  }).then(async () => {
    try {
      await deleteTag(ids)
      ElMessage.success('删除成功')
      proTable.value?.clearSelection()
      proTable.value?.getTableList()
    } catch (error) {
      console.log(error)
      ElMessage.error('删除失败')
    }
  })
}
</script>
```

点击"删除及批量删除",页面效果如图5,图6所示。

图5　删除确认图

图 6　批量删除确认图

3　任务总结

本次任务成功完成了后台管理系统中标签管理模块的开发,涵盖了从后端模型设计到前端页面展示的全流程。

在后端部分,我们注重数据模型的设计、状态的枚举化、数据传输对象的标准化、数据转换的便捷性、查询模型的灵活性、数据库交互的高效性、业务逻辑的完善性以及 API 接口的清晰性。通过创建 Tag 实体类、CommonStatusEnum 枚举类、TagEditDTO 和 TagVO 数据传输对象,我们确保了数据的准确性和前后端交互的标准化。同时,通过数据转换接口、查询模型、Mapper 层以及 Service 层的实现,我们提供了灵活的查询功能和完善的业务逻辑。最后,在 Controller 层定义了清晰的 API 接口,为前端提供了稳定的服务支持。

在前端部分,我们聚焦于接口的定义、API 的封装、页面的修改、操作模式的判断以及删除功能的实现。通过定义类型声明和封装 API 接口,我们确保了前后端数据交互的准确性和效率。在页面展示上,我们根据需求修改了标签的展示页面,并实现了新增、编辑、查看标签的功能,提升了用户体验。同时,我们通过判断操作模式和添加自定义作用域插槽,支持了批量删除和单个删除功能,并确保了删除操作后页面数据的实时更新。

任务八

分类管理模块开发

波哥:"我们今天要讨论并分配一个新任务——分类管理模块的开发。这个模块是后台管理系统的一个重要部分,主要用于管理各种分类信息的增、删、改、查。具体来说,就是要在数据库中设计合理的表结构来存储分类信息,然后创建相应的后端接口来处理前端发送的增、删、改、查请求。前端部分你们需要设计一个用户友好的界面来展示分类信息,并提供增、删、改、查的操作按钮。当用户点击这些按钮时,前端需要发送相应的请求到后端,并处理后端返回的数据来更新界面。现在就开始吧!有任何问题及时沟通。"

◇ 任务点

- 分类管理模块的后端实现;
- 分类管理模块的前端实现。

◇ 任务计划

- 任务内容:分类管理模块的开发;
- 任务耗时:预计完成时间为 1 h;
- 任务难点:无。

1 分类管理模块后端实现

1.1 任务描述

在波哥的指导下,小南和小工将携手负责分类管理模块的后端实现。

首先需要定义与数据库 t_category 表对应的 Category 数据模型,确保分类数据在系统中的准确存储。小南将负责创建 CategoryEditDTO 类,用于接收和封装前端发送的分类编辑请求数据,并设计 CategoryVO 类,封装返回给前端的分类数据,以满足前端的展示需求。

小工承担数据转换的工作,在 convert 包中实现 Category、CategoryEditDTO 和 CategoryVO 之间的数据转换逻辑,确保数据在不同层之间准确传递。

在查询功能方面,小南将定义 CategoryQuery 类,封装分类查询的条件,以支持灵活的查询操作。而小工则会在 mapper 包中定义 CategoryMapper 接口,并编写对应的 SQL 映射文件,实现分类数据的增、删、改、查操作。

在业务逻辑处理层面,小工将在 service 包中定义 CategoryService 接口,并在 service.impl 包中实现具体的业务逻辑,包括数据验证、事务管理等。

在 controller 包中,小工将创建 CategoryController 类,定义分类管理的 API 接口,并调用 Service 层实现具体的业务功能。

1.2 任务分析

1.2.1 PO 实体创建

在 model.entity 包下新建 Category 类,对应 t_category 表。

```
@Data
@TableName("t_category")
public class Category {

    @TableId(value = "pk_id", type = IdType.AUTO)
    private Integer pkId;
    private String title;
    /**
     * 0-网盘,1-资源
     */
    private Integer type;
    private String description;
    @TableField(value = "delete_flag", fill = FieldFill.INSERT)
    @TableLogic
    private Integer deleteFlag;
    @TableField(value = "update_time", fill = FieldFill.INSERT_UPDATE)
    private LocalDateTime updateTime;
    @TableField(value = "create_time", fill = FieldFill.INSERT)
    private LocalDateTime createTime;

}
```

1.2.2 DTO 实体创建

在 model.dto 包下新建 CategoryEditDTO 类,用于分类对象信息的新增和修改。

```
@Data
@Schema(name = "CategoryEditDTO", description = "分类编辑 DTO")
public class CategoryEditDTO {
```

@Schema(name = "pkId", description = "主键")
private Integer pkId;
@Schema(name = "title", description = "标题")
private String title;
@Schema(name = "type", description = "类型")
private Integer type;
@Schema(name = "description", description = "描述")
private String description;
@Schema(name = "deleteFlag", description = "删除标识")
private Integer deleteFlag;
}

1.2.3 VO 返回实体创建

在 model.vo 包中新建 CategoryVO 类,用于返回给前端的分类视图对象。

```
@Data
@Schema(name = "CategoryVO", description = "分类返回 vo")
public class CategoryVO {
    @Schema(name = "pkId", description = "主键")
    private Integer pkId;
    @Schema(name = "title", description = "标题")
    private String title;
    @Schema(name = "type", description = "类型")
    private Integer type;
    @Schema(name = "description", description = "描述")
    private String description;
    @Schema(name = "createTime", description = "创建时间")
    @JsonFormat(pattern = "yyyy-MM-dd HH:mm:ss", timezone = "GMT+8")
    private LocalDateTime createTime;
}
```

1.2.4 Convert 转换创建

在 convert 包中新建 CategoryConvert 转换接口,用于分类相关的实体类转换。

```
@Mapper
public interface CategoryConvert {
    CategoryConvert INSTANCE = Mappers.getMapper(CategoryConvert.class);

    Category convert(CategoryEditDTO dto);
}
```

1.2.5 Query 查询实体创建

在 model.query 包中新建 CategoryQuery。继承 Query 查询基类,用于条件查询。

```java
@EqualsAndHashCode(callSuper = true)
@Data
@Schema(name = "CategoryQuery", description = "分类查询")
public class CategoryQuery extends Query {
    @Schema(name = "标题")
    private String title;

    @Schema(name = "类型")
    private Integer type;
}
```

1.2.6 Mapper 层实现

1. 在 mapper 包中创建 CategoryMapper 接口,定义分页查询方法。

```java
public interface CategoryMapper extends BaseMapper<Category> {

    List<CategoryVO> getCategoryPage(Page<CategoryVO> page, @Param("query") CategoryQuery query);

}
```

2. 在 resources/mapper 下创建 CategoryMapper.xml,实现分页方法。

```xml
<?xml version="1.0" encoding="UTF-8"?>
<!DOCTYPE mapper PUBLIC "-//mybatis.org//DTD Mapper 3.0//EN" "http://mybatis.org/dtd/mybatis-3-mapper.dtd">
<mapper namespace="top.ssy.share.admin.mapper.CategoryMapper">

    <select id="getCategoryPage" resultType="top.ssy.share.admin.model.vo.CategoryVO">
        SELECT tc.* FROM t_category tc
        WHERE tc.delete_flag = 0
        <if test="query.title != null and query.title != '' ">
            AND tc.title LIKE concat('%',#{query.title},'%')
        </if>
        <if test="query.type != null ">
            AND tc.type = #{query.type}
```

```
        </if>
        ORDER BY tc.create_time DESC
    </select>

</mapper>
```

1.2.7 Service 层业务实现

1. 在 service 包中定义 CategoryService 接口,实现基础的增、删、改、查接口。

```
public interface CategoryService extends IService<Category> {

    PageResult<CategoryVO> page(CategoryQuery query);

    void saveAndEdit(CategoryEditDTO dto);

    void delete(List<Integer> id);
}
```

2. 在 service.impl 包中新建 CategoryServiceImpl 类,实现接口方法。

```
@Service
public class CategoryServiceImpl extends ServiceImpl<CategoryMapper, Category> implements CategoryService {

    @Override
    public PageResult<CategoryVO> page(CategoryQuery query) {
        Page<CategoryVO> page = new Page<>(query.getPage(), query.getLimit());
        List<CategoryVO> list = baseMapper.getCategoryPage(page, query);
        return new PageResult<>(list, page.getTotal());
    }

    @Override
    public void saveAndEdit(CategoryEditDTO dto) {
        Category category = CategoryConvert.INSTANCE.convert(dto);
        if (dto.getPkId() == null) {
            save(category);
        } else {
            updateById(category);
        }
    }
```

```
        }

        @Override
        public void delete(List<Integer> id) {
            baseMapper.deleteBatchIds(id);
        }
    }
```

1.2.8　Controller 层接口实现

在 controller 层中新建 CategoryController 接口,定义 API 接口方法。

```
package top.ssy.share.admin.controller;

import io.swagger.v3.oas.annotations.Operation;
import io.swagger.v3.oas.annotations.tags.Tag;
import jakarta.validation.Valid;
import lombok.AllArgsConstructor;
import org.springframework.security.access.prepost.PreAuthorize;
import org.springframework.web.bind.annotation.PostMapping;
import org.springframework.web.bind.annotation.RequestBody;
import org.springframework.web.bind.annotation.RequestMapping;
import org.springframework.web.bind.annotation.RestController;
import top.ssy.share.admin.common.result.PageResult;
import top.ssy.share.admin.common.result.Result;
import top.ssy.share.admin.model.dto.CategoryEditDTO;
import top.ssy.share.admin.model.query.CategoryQuery;
import top.ssy.share.admin.model.vo.CategoryVO;
import top.ssy.share.admin.service.CategoryService;

import java.util.List;


@RestController
@AllArgsConstructor
@Tag(name = "分类管理", description = "分类管理")
@RequestMapping("/category")
public class CategoryController {
    private final CategoryService categoryService;
```

```java
@PostMapping("/page")
@Operation(summary = "分页")
@PreAuthorize("hasAuthority('sys:category:view')")
public Result<PageResult<CategoryVO>> page(@RequestBody @Valid CategoryQuery query) {
    return Result.ok(categoryService.page(query));
}

@PostMapping("saveAndEdit")
@Operation(summary = "新增或修改")
@PreAuthorize("hasAnyAuthority('sys:category:add', 'sys:category:edit')")
public Result<String> savedAndEdit(@RequestBody @Valid CategoryEditDTO dto) {
    categoryService.saveAndEdit(dto);
    return Result.ok();
}

@PostMapping("/delete")
@Operation(summary = "删除")
@PreAuthorize("hasAuthority('sys:category:remove')")
public Result<String> delete(@RequestBody List<Integer> ids) {
    categoryService.delete(ids);
    return Result.ok();
}
}
```

2 分类管理模块前端实现

2.1 任务描述

在分类管理模块的前端实现过程中,小南和小工需要共同协作完成一系列关键任务。

小南首先在 src/api/interface/index.ts 文件中定义与分类管理相关的类型声明,确保前端代码能够准确理解后端 API 返回的数据结构和类型。同时在 src/api/modules/category/index.ts 文件中新增封装分类管理相关接口的代码,使得前端能够方便地与后端进行通信,实现分类数据的获取和发送。

小工负责编写 src/views/Resource/Category.vue 文件,实现分类的展示页面。参考之前的开发经验,设计一个直观易用的分类列表界面,展示从后端获取的分类数据。为了实现分类的编辑和查看功能,还需要修改分类弹框组件 src/views/Resource/components/CategoryDialog.vue。

为了支持分类的删除操作,小工需要在 Category.vue 页面中添加删除相关的代码。编写删除按钮的点击事件处理函数,调用后端 API 接口实现分类的删除,并更新页面数据以反映删除后的结果。

2.2 任务分析

2.2.1 定义类型声明

在 src/api/interface/index.ts 中添加与分类管理相关的类型声明。

```
/** 分类管理 */
export namespace SysCategory {
  export interface ReqGetCategoryParams extends ReqPage {
    name?: string
  }
  /** 分类列表 */
  export interface ResCategoryList {
    pkId: number
    title: string
    type: number
    description: string
    deleteFlag: number
    createTime: string
    updateTime: string
  }
  /** 分类编辑 */
  export interface ReqEditCategoryParams {
    title: string
    type: number
    description: string
  }
  /** 新增分类 */
  export interface ReqAddCategoryParams {
    title: string
    type: number
    description: string
  }
}
```

2.2.2 添加接口

在 src/api/modules/category/index.ts 中新增封装分类管理相关接口。

```
import { ResPage, SysCategory } from '@/api/interface/index'
import { _API } from '@/api/axios/servicePort'
import http from '@/api'

// 分类列表
export const getCategoryPage = (params: SysCategory.ReqGetCategoryParams) => {
  return http.post<ResPage<SysCategory.ResCategoryList>>(_API + '/category/page', params)
}

// 添加分类
export const addCategory = (params: SysCategory.ReqAddCategoryParams) => {
  return http.post(_API + '/category/saveAndEdit', params)
}

// 编辑分类
export const editCategory = (params: SysCategory.ReqEditCategoryParams) => {
  return http.post(_API + '/category/saveAndEdit', params)
}

// 删除分类
export const deleteCategory = (params: number[]) => {
  return http.post(_API + '/category/delete', params)
}
```

2.2.3 分类展示页面

编写 src/views/Resource/Category.vue，实现分类的展示页面。

```
<template>
  <div class="table-box">
    <ProTable ref="proTable" title="分类列表" :columns="columns" :requestApi="getTableList" :initParam="initParam" :dataCallback="dataCallback">
      <!-- 表格 header 按钮 -->
      <template #tableHeader>
        <el-button type="primary" :icon="CirclePlus" @click="openDrawer('新增')" v-hasPermi="['sys:category:add']">新增分类</el-button>
      </template>
      <!-- 表格操作 -->
```

```vue
          <template #operation="scope">
            <el-button type="primary" link :icon="View" @click="openDrawer('查看', scope.row)" v-hasPermi="['sys:category:view']">查看</el-button>
            <el-button type="primary" link :icon="EditPen" @click="openDrawer('编辑', scope.row)" v-hasPermi="['sys:category:edit']">编辑</el-button>
          </template>
    </ProTable>
    <CategoryDialog ref="dialogRef" />
  </div>
</template>

<script setup name="SysCategory" lang="tsx">
import { SysCategory } from '@/api/interface'
import { ref, reactive } from 'vue'
import { CirclePlus, EditPen, View } from '@element-plus/icons-vue'
import { ColumnProps } from '@/components/ProTable/interface'
import ProTable from '@/components/ProTable/index.vue'
import CategoryDialog from '@/views/Resource/components/CategoryDialog.vue'
import { getCategoryPage, addCategory, editCategory } from '@/api/modules/category'
import { dictConfigList } from '@/api/modules/dict/dictConfig'

// 获取 ProTable 元素,调用其获取刷新数据方法(还能获取当前查询参数,方便导出携带参数)
const proTable = ref()

// 如果表格需要初始化请求参数,直接定义传给 ProTable(之后每次请求都会自动带上该参数,此参数
// 更改之后也会一直带上,改变此参数会自动刷新表格数据)
const initParam = reactive({})

// dataCallback 是对于返回的表格数据做处理,如果你后台返回的数据不是 datalist && total 这些字段,
// 那么你可以在这里处理成这些字段
const dataCallback = (data: any) => {
  return {
    list: data.list,
    total: data.total
  }
}
```

```
// 如果你想在请求之前对当前请求参数做一些操作,可以自定义如下函数:params 为当前所有的请求参
// 数(包括分页),最后返回请求列表接口
// 若默认不做操作就直接在 ProTable 组件上绑定 :requestApi = "getUserList"
const getTableList = (params: any) => {
  let newParams = { ...params }
  return getCategoryPage(newParams)
}


// 表格配置项
const columns: ColumnProps[] = [
  { type: 'selection', fixed: 'left', width: 60 },
  {
    prop: 'title',
    label: '分类标题',
    showOverflowTooltip: true,
    search: { el: 'input' }
  },
  {
    prop: 'type',
    label: '分类',
    width: 100,
    search: { el: 'select', props: { filterable: true } },
    // enum: [
    //   { title: '网盘类型', value: 0 },
    //   { title: '资源类型', value: 1 }
    // ],
    enum: () => dictConfigList('categoryType'),
    fieldNames: { label: 'title', value: 'value' },
    render: (scope) => {
      return <el-tag type={scope.row.type === 0 ? 'success' : 'primary'}>{scope.row.type == 0 ? '网盘类型' : '资源类型'}</el-tag>
    }
  },
  {
    prop: 'description',
    label: '分类描述',
    showOverflowTooltip: true
  },
```

```
      {
        prop: 'createTime',
        label: '创建时间',
        width: 200
      },
      { prop: 'operation', label: '操作', fixed: 'right', width: 330 }
    ]

    // 打开 drawer(新增、查看、编辑)
    const dialogRef = ref()
    const openDrawer = (title: string, row: Partial<SysCategory.ResCategoryList> = {}) => {
      let params = {
        title,
        row: { ...row },
        isView: title === '查看',
        api: title === '新增' ? addCategory : title === '编辑' ? editCategory : '',
        getTableList: proTable.value.getTableList,
        maxHeight: '200px'
      }
      dialogRef.value.acceptParams(params)
    }
</script>
```

获取后端数据,页面效果如图1所示。

图1　分类展示页面效果图

2.2.4 新增编辑查看功能

修改分类弹框组件 src/views/Resource/components/CategoryDialog.vue。

```
<template>
  <Dialog :model-value="dialogVisible" :title="dialogProps.title" :fullscreen="dialogProps.fullscreen" :max-height="dialogProps.maxHeight" :cancel-dialog="cancelDialog">
    <div :style="'width: calc(100% - ' + dialogProps.labelWidth! / 2 + 'px)'">
      <el-form
        ref="ruleFormRef"
        label-position="right"
        :label-width="dialogProps.labelWidth + 'px'"
        :rules="rules"
        :model="dialogProps.row"
        :disabled="dialogProps.isView"
        :hide-required-asterisk="dialogProps.isView"
      >
        <el-form-item label="分类名称" prop="title">
          <el-input v-model="dialogProps.row!.title" placeholder="请填写分类名称(10字以内)" clearable></el-input>
        </el-form-item>
        <el-form-item label="分类类别" prop="type">
          <!-- select -->
          <el-select v-model="dialogProps.row!.type" placeholder="请选择分类类别" clearable>
            <el-option label="网盘类型" :value="'0'"></el-option>
            <el-option label="资源类型" :value="'1'"></el-option>
          </el-select>
        </el-form-item>
        <el-form-item label="分类描述" prop="description">
          <el-input v-model="dialogProps.row!.description" placeholder="请填写分类描述(200字以内)" :rows="3" type="textarea" maxlength="200"></el-input>
        </el-form-item>
      </el-form>
    </div>
    <template #footer>
      <slot name="footer">
        <el-button @click="cancelDialog">取消</el-button>
```

```vue
            <el-button type="primary" v-show="!dialogProps.isView" @click="handleSubmit">确定</el-button>
        </slot>
      </template>
    </Dialog>
</template>

<script setup lang="ts">
import { ref, reactive } from 'vue'
import { ElMessage, FormInstance } from 'element-plus'
import { Dialog } from '@/components/Dialog'

interface DialogProps {
  title: string
  isView: boolean
  fullscreen?: boolean
  row: any
  labelWidth?: number
  maxHeight?: number | string
  api?: (params: any) => Promise<any>
  getTableList?: () => Promise<any>
  treeMenuList?: any
}
const dialogVisible = ref(false)
const dialogProps = ref<DialogProps>({
  isView: false,
  title: '',
  row: {},
  labelWidth: 160,
  fullscreen: true,
  maxHeight: '500px'
})

// 接收父组件传过来的参数
const treeStrictly = ref(true)
const acceptParams = (params: DialogProps): void => {
  params.row = { ...dialogProps.value.row, ...params.row }
  dialogProps.value = { ...dialogProps.value, ...params }
```

```
    dialogVisible.value = true
}

defineExpose({
    acceptParams
})

const rules = reactive({
    title: [
        { required: true, message: '请输入分类标题', trigger: 'blur' },
        {
            min: 2,
            max: 10,
            message: '长度在 2 到 10 个字符',
            trigger: 'blur'
        }
    ],
    description: [
        { required: true, message: '请输入分类描述', trigger: 'blur' },
        {
            min: 1,
            max: 200,
            message: '长度在 1 到 200 个字符',
            trigger: 'blur'
        }
    ],
    type: [{ required: true, message: '请选择分类类型', trigger: 'blur' }]
})
const ruleFormRef = ref<FormInstance>()
const handleSubmit = () => {
    ruleFormRef.value!.validate(async (valid) => {
        if (!valid) return
        try {
            await dialogProps.value.api!(dialogProps.value.row)
            ElMessage.success({ message: `${dialogProps.value.title}分类成功！` })
            dialogProps.value.getTableList!()
            cancelDialog(true)
        } catch (error) {
```

```
        console.log(error)
      }
    })
  }
  const cancelDialog = (isClean?: boolean) => {
    dialogVisible.value = false
    let condition = ['查看', '编辑']
    if (condition.includes(dialogProps.value.title) || isClean) {
      dialogProps.value.row = {}
      ruleFormRef.value!.resetFields()
    }
    treeStrictly.value = true
  }
</script>

<style scoped lang="less">
  :deep(.penultimate-node) {
    .el-tree-node__children {
      padding-left: 60px;
      line-height: 12px;
      white-space: pre-wrap;

      .el-tree-node {
        display: inline-block;
      }

      .el-tree-node__content {
        padding-right: 5px;
        padding-left: 5px !important;

        .el-tree-node__expand-icon {
          display: none;
        }
      }
    }
  }
</style>
```

点击"新增",页面效果如图 2 所示。

图 2 "新增"页面效果图

点击"编辑",页面效果如图 3 所示。

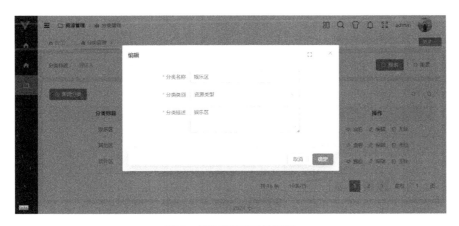

图 3 "编辑"页面效果图

点击"查看"按钮,页面效果如图 4 所示。

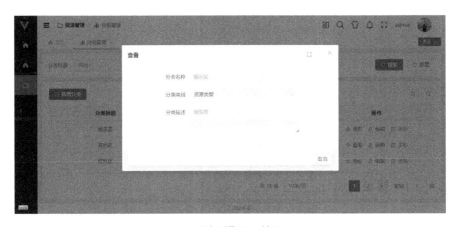

图 4 "查看"页面效果图

2.2.5 删除功能

修改 Category.vue 页面,添加删除相关的代码。

```vue
<template>
  <div class="table-box">
    <ProTable ref="proTable" title="分类列表" :columns="columns" :requestApi="getTableList" :initParam="initParam" :dataCallback="dataCallback">
      <!-- 表格 header 按钮 -->
      <template #tableHeader>
        <el-button type="primary" :icon="CirclePlus" @click="openDrawer('新增')" v-hasPermi="['sys:category:add']">新增分类</el-button>
      </template>
      <!-- 表格操作 -->
      <template #operation="scope">
        <el-button type="danger" link :icon="Delete" @click="deleteRow(scope.row)" v-hasPermi="['sys:category:remove']">删除</el-button>
      </template>
    </ProTable>
    <CategoryDialog ref="dialogRef" />
  </div>
</template>
<script setup name="SysCategory" lang="tsx">
import { Delete } from '@element-plus/icons-vue'
import { deleteCategory } from '@/api/modules/category'
import { useHandleData } from '@/hooks/useHandleData'

// 删除单个
const deleteRow = async (params: SysCategory.ResCategoryList) => {
  await useHandleData(deleteCategory, [params.pkId], `删除【${params.title}】分类`)
  proTable.value.getTableList()
}
</script>
```

点击"删除",页面效果如图 5、图 6 所示。

图 5　"删除"操作确认页面

图 6　删除成功页面效果

3　任务总结

在分类管理模块的开发过程中,我们分别从后端和前端两个维度完成了关键任务,确保了模块的完整性和功能性。

在后端实现上,我们注重于数据模型的定义、数据传输对象的创建、视图对象的构建、实体转换的实现、查询条件的封装、数据库交互的完成以及业务逻辑和 API 接口的处理。具体而言,我们定义了 Category 实体类与数据库表对应,确保了数据的持久化;创建了 CategoryEditDTO 和 CategoryVO,分别用于前后端的数据传输和展示;实现了 CategoryConvert

资源分享应用后台管理系统开发实战

接口,简化了数据转换过程;封装了 CategoryQuery 类,使得查询条件更加清晰;通过 CategoryMapper 接口及 XML 映射文件,实现了与数据库的交互;在 Service 层完成了 CategoryService 接口及其实现类,处理了业务逻辑;最后,在 Controller 层定义了 API 接口,为前端提供了服务支持。

在前端实现上,我们聚焦于类型声明的定义、API 接口的封装、页面展示的开发以及交互功能的实现。我们首先在 index.ts 中定义了与分类管理相关的类型声明,确保了数据类型的准确性;接着在 category/index.ts 中封装了 API 接口,为前端与后端的通信提供了桥梁;然后实现了 Category.vue 页面,用于分类的展示;开发了 CategoryDialog.vue 组件,支持了分类的编辑与查看;最后,在 Category.vue 页面中添加了删除功能,并通过调用后端 API 实现了删除操作。

任务九

资源管理模块开发

波哥在团队群里发布了新的任务："我们接下来要着手做资源管理模块了。资源管理，简单来说就是让管理员查看目前存在的资源。同时，我们还需要一个审核功能，确保上传的资源符合我们的要求。"

"我们首先要设计数据库来存储资源信息，然后开发后端接口来处理资源的查看及审核。前端方面，要设计一个简洁易用的界面让用户操作。最后，别忘了实现审核功能。"

◇ 任务点

- 资源列表的后端和前端实现；
- 审核资源的后端和前端实现；
- 积分管理模块的补充业务实现。

◇ 任务计划

- 任务内容：完成后台管理系统中资源管理模块的开发；
- 任务耗时：预计完成时间为 1.5 h；
- 任务难点：审核资源功能的实现。

1 资源列表

1.1 任务描述

为了完善资源列表功能，波哥、小南和小工将分别承担不同的任务来推动该功能的实现。

波哥负责后端的基础类创建。在 entity 包中新建一个 Resource 实体类，确保该类能够与数据库中的 t_resource 表对应，以便进行数据操作。在 enums 包中创建一个名为 ResourceStatusEnum 的枚举类，用于定义资源的审核状态，为资源的状态管理提供基础。此外，为了支持复杂的查询操作，波哥还计划在 query 包中新建一个 ResourceQuery 查询实体。

数据视图也由波哥负责。应在 vo 包中新建一个 ResourceVO 类，用于作为接口返回视图实体，确保前端能够获取所需的数据格式。为了简化数据转换过程，他将在 convert 包中定义一个 ResourceConvert 实体转换接口，用于在 Resource 实体和 ResourceVO 视图之间进行数据转换。最后,波哥还将在 mapper 包中新建一个 ResourceMapper 接口,用于数据库层的交互。

小南负责 Service 层的业务实现。首先在 service 包中新建一个 ResourceService 接口,并定义一个分页查询方法。接着,在 service.impl 包中创建一个 ResourceServiceImpl 类来实现这个接口方法。虽然在这个过程中可能会遇到尚未实现的方法导致报错,但小南已经做好了后续实现对应方法的准备。

小工则专注于前端的工作。首先在后台管理系统前端项目的 src/api/interface/index.ts 文件中定义与资源类型相关的声明,确保前端能够正确地识别和使用这些数据类型。接着,在 src/api/modules/resource/index.ts 文件中封装资源列表和审核资源的接口,为前端提供与后端通信的桥梁。此外,为了支持资源审核功能,他还计划在 src/api/modules/user/index.ts 文件中新增一个查询所有用户的接口。最后,小工将编辑资源组件 src/views/Resource/Resource.vue,搭建资源列表的相关页面。他将配置好对应的表格属性,确保页面能够正确地展示从后端获取的资源数据。

1.2 任务分析

1.2.1 基础类创建

1.2.1.1 PO 实体创建

在 entity 包中新建 Resource 实体类,对应 t_resource 表。

```
@Data
@TableName(value = "t_resource", autoResultMap = true)
public class Resource {

    @TableId(value = "pk_id", type = IdType.AUTO)
    private Integer pkId;
    private String title;
    private Integer author;
    private Integer diskType;
    @TableField(typeHandler = FastjsonTypeHandler.class)
    private List<Integer> resType;
    @TableField(typeHandler = FastjsonTypeHandler.class)
    private List<Integer> tags;
    private String downloadUrl;
    private String detail;
    private Integer price;
    private Integer likeNum;
    /**
```

```
     * @see top.ssy.share.admin.enums.CommonStatusEnum
     */
    private Integer isTop;
    /**
     * @see top.ssy.share.admin.enums.ResourceStatusEnum
     */
    private Integer status;
    private String remark;
    @TableField(value = "delete_flag", fill = FieldFill.INSERT)
    @TableLogic
    private Integer deleteFlag;
    @TableField(value = "update_time", fill = FieldFill.INSERT_UPDATE)
    private LocalDateTime updateTime;
    @TableField(value = "create_time", fill = FieldFill.INSERT)
    private LocalDateTime createTime;

}
```

1.2.1.2 ResourceStatusEnum 资源状态枚举

在 enums 包中新建 ResourceStatusEnum 枚举类,定义资源的审核状态。

```
@Getter
public enum ResourceStatusEnum {

    /**
     * 未审核
     */
    UNAUDITED(0, "未审核"),
    /**
     * 审核通过
     */
    AUDITED(1, "审核通过"),
    /**
     * 审核不通过
     */
    NOT_AUDITED(2, "审核不通过")
    ;

    private final Integer code;
```

```
    private final String status;

    ResourceStatusEnum(Integer code, String status) {
        this.code = code;
        this.status = status;
    }
}
```

1.2.1.3 Query 查询实体

在 query 包中新建 ResourceQuery 查询实体。

```
@Data
@EqualsAndHashCode(callSuper = true)
@Schema(name = "resourceQuery", description = "资源查询")
public class ResourceQuery extends Query {
    @Schema(name = "title", description = "标题")
    private String title;
    @Schema(name = "author", description = "作者")
    private Integer author;
    @Schema(name = "isTop", description = "是否置顶")
    private Integer isTop;
    @Schema(name = "status", description = "状态")
    private Integer status;
}
```

1.2.1.4 VO 返回视图创建

在 vo 包中新建 ResourceVO,用作接口返回视图实体。

```
@Data
@Schema(name = "ResourceVO", description = "资源返回 vo")
public class ResourceVO {
    @Schema(name = "pkId", description = "主键")
    private Integer pkId;
    @Schema(name = "title", description = "标题")
    private String title;
    @Schema(name = "authorName", description = "作者 id")
    private Integer author;
    @Schema(name = "authorName", description = "作者")
    private String authorName;
```

@Schema(name = "diskType", description = "网盘分类")
private String diskType;
@Schema(name = "resType", description = "资源分类")
private List<Integer> resType;
@Schema(name = "tags", description = "标签列表")
private List<Integer> tags;
@Schema(name = "resTypeList", description = "资源分类列表")
private List<String> resTypeList;
@Schema(name = "tagList", description = "标签列表")
private List<String> tagList;
@Schema(name = "downloadUrl", description = "下载资源")
private String downloadUrl;
@Schema(name = "detail", description = "详情")
private String detail;
@Schema(name = "price", description = "价格")
private Integer price;
@Schema(name = "likeNum", description = "点赞数")
private Integer likeNum;
@Schema(name = "isTop", description = "是否置顶")
private Integer isTop;
@Schema(name = "status", description = "审核状态")
private Integer status;
@Schema(name = "remark", description = "审核描述")
private String remark;
@Schema(name = "createTime", description = "创建时间")
@JsonFormat(pattern = "yyyy-MM-dd HH:mm:ss", timezone = "GMT+8")
private LocalDateTime createTime;
}

1.2.1.5 Converter 实体转换

在 convert 包中新建 ResourceConvert 实体转换接口。

```
@Mapper
public interface ResourceConvert {
    ResourceConvert INSTANCE = Mappers.getMapper(ResourceConvert.class);

    ResourceVO convert(Resource resource);
}
```

1.2.1.6 Mapper 层创建

在 mapper 包中新建 ResourceMapper 接口。

```java
public interface ResourceMapper extends BaseMapper<Resource> {

}
```

1.2.2 Service 层业务实现

1. 在 service 包中新建 ResourceService 接口,创建分页查询方法。

```java
public interface ResourceService extends IService<Resource> {

    /**
     * 分页
     *
     * @param query 查询
     * @return {@link PageResult}<{@link ResourceVO}>
     */
    PageResult<ResourceVO> page(ResourceQuery query);
}
```

2. 在 service.impl 包中新建 ResourceServiceImpl 类,实现接口方法。代码中会有两个方法报错,因为还没实现,接下来马上会进行实现。

```java
@Slf4j
@Service
@AllArgsConstructor
public class ResourceServiceImpl extends ServiceImpl<ResourceMapper, Resource> implements ResourceService {
    private final TagService tagService;
    private final CategoryService categoryService;
    private final UserService userService;

    @Override
    public PageResult<ResourceVO> page(ResourceQuery query) {
        LambdaQueryWrapper<Resource> wrapper = new LambdaQueryWrapper<>();
        wrapper
                .like(StringUtils.isNotBlank(query.getTitle()), Resource::getTitle, "%" + query.getTitle() + "%")
                .eq(ObjectUtils.isNotEmpty(query.getAuthor()), Resource::getAuthor, query.getAuthor())
```

```
                .eq(ObjectUtils.isNotEmpty(query.getIsTop()), Resource::getIsTop, query.getIsTop())
                .eq(ObjectUtils.isNotEmpty(query.getStatus()), Resource::getStatus, query.getStatus());
        Page<Resource> resourcePage = baseMapper.selectPage(new Page<>(query.getPage(),
query.getLimit()), wrapper);
        List<Resource> list = resourcePage.getRecords();
        List<ResourceVO> resourceVOList = list.stream().map(item -> {
            ResourceVO resource = ResourceConvert.INSTANCE.convert(item);
            Category diskType = categoryService.getById(item.getDiskType());
            User author = userService.getById(item.getAuthor());
            resource.setAuthorName(author.getNickname());
            resource.setDiskType(diskType.getTitle());

resource.setResTypeList(categoryService.listByPkIdList(resource.getResType()));
            resource.setTagList(tagService.listByPkIdList(resource.getTags()));
            return resource;
        }).collect(Collectors.toList());
        return new PageResult<>(resourceVOList, resourcePage.getTotal());
    }
}
```

1.2.3 补充方法

1.2.3.1 DeleteFlagEnum 删除标识枚举

在 enums 包中新建 DeleteFlagEnum 枚举。

```
@Getter
public enum DeleteFlagEnum {
    /**
     * 未删除
     */
    NOT_DELETE(0, "未删除"),
    /**
     * 已删除
     */
    DELETED(1, "已删除");

    private final Integer code;
    private final String desc;
```

```
DeleteFlagEnum(Integer code, String desc) {
    this.code = code;
    this.desc = desc;
}
```
}

1.2.3.2 Service 层方法创建

1. CategoryService 接口新增方法,根据分类的主键集合查询分类的名字。

```
List<String> listByPkIdList(List<Integer> pkIdList);
```

2. TagService 接口新增方法,根据标签的主键集合查询标签的名字。

```
List<String> listByPkIdList(List<Integer> pkIdList);
```

1.2.3.3 Service 方法实现

1. CategoryServiceImpl 实现新建的方法。

```
@Override
public List<String> listByPkIdList(List<Integer> pkIdList) {
    return baseMapper.selectBatchIds(pkIdList)
            .stream()
            .filter(category -> Objects.equals(category.getDeleteFlag(),
DeleteFlagEnum.NOT_DELETE.getCode()))
            .map(Category::getTitle)
            .collect(Collectors.toList());
}
```

2. TagServiceImpl 实现新建的方法。

```
@Override
public List<String> listByPkIdList(List<Integer> pkIdList) {
    return baseMapper.selectBatchIds(pkIdList)
            .stream()
            .filter(tag -> Objects.equals(tag.getDeleteFlag(),
DeleteFlagEnum.NOT_DELETE.getCode()))
            .map(Tag::getTitle)
            .collect(Collectors.toList());
}
```

两个方法主要是根据表中的主键集合,批量查询对应的实体,筛选出未删除的数据,将它们的标题再收集成集合,方便使用展示。

1.2.4 Controller 层接口实现

在 controller 包中新建 ResourceController 类,实现分页查询接口。

```
@RestController
@AllArgsConstructor
@RequestMapping("/resource")
@Tag(name = "资源管理", description = "资源管理")
public class ResourceController {
    private final ResourceService resourceService;

    @PostMapping("/page")
    @Operation(summary = "分页")
    @PreAuthorize("hasAuthority('sys:resource:view')")
    public Result<PageResult<ResourceVO>> page(@RequestBody @Valid ResourceQuery query) {
        return Result.ok(resourceService.page(query));
    }
}
```

1.2.5 前端定义类型及接口

后台管理系统前端项目,在 src/api/interface/index.ts 中定义与资源类型相关声明。

```
/** 资源类型 */
export namespace SysResource {
  /** 资源请求参数 */
  export interface ReqGetResourceParams extends ReqPage {
    name?: string
  }
  /** 资源列表 */
  export interface ResResourceList {
    pkId: number
    title: string
    author: number
    diskType: number
    resType: number
    tags: string[]
    downloadUrl: string
    detail: string
```

```
      price: number
      likeNum: number
      isTop: number
      status: number
      remark: string
      deleteFlag: number
      createTime: string
      updateTime: string
      tagList: string[]
      resTypeList: string[]
    }
}
```

新建 src/api/modules/resource/index.ts，封装资源列表、审核资源接口。

```
import { ResPage, SysResource } from '@/api/interface/index'
import { _API } from '@/api/axios/servicePort'
import http from '@/api'


// 资源列表
export const getResourcePage = (params: SysResource.ReqGetResourceParams) => {
  return http.post<ResPage<SysResource.ResResourceList>>(_API + '/resource/page', params)
}

// 审核资源
export const auditResource = (params: { pkId: number; status: number; remark: string }) => {
  return http.post(_API + '/resource/audit', params)
}
```

在 src/api/modules/user/index.ts 中，新增查询所有用户的接口。

```
import http from '@/api'
import { _API } from '@/api/axios/servicePort'

/**
 * @name 用户管理模块
 */
export const UserApi = {
  // 查询用户列表
```

```
    page: (params: any) => http.post(_API + '/user/page', params),
    // 编辑用户
    edit: (params: any) => http.post(_API + '/user/edit', params),
    // 导出用户列表
    export: (params: any) =>
      http.post(_API + '/user/export', params, {
        responseType: 'blob'
      }),
    // 冻结用户
    freezeUser: (userId: number) => http.post(_API + '/user/enabled? userId=' + userId),
    // 查询用户
    findAllUser: () => http.get<UserType[]>(_API + '/user/list')
}
```

1.2.6 搭建展示列表

编辑资源组件 src/views/Resource/Resource.vue,搭建资源列表的相关页面,配置好对应的表格属性,就可以查看到页面的效果。

```
<template>
  <div class="table-box">
    <ProTable ref="proTable" title="资源列表" :columns="columns" :requestApi="getTableList" :initParam="initParam" :dataCallback="dataCallback"> </ProTable>
  </div>
</template>

<script setup lang="tsx" name="SysResource">
import { ref, reactive, onMounted } from 'vue'
import { SysResource } from '@/api/interface'
import { ColumnProps } from '@/components/ProTable/interface'
import ProTable from '@/components/ProTable/index.vue'
import { getResourcePage } from '@/api/modules/resource'
import { UserApi } from '@/api/modules/user'
import { dictConfigList } from '@/api/modules/dict/dictConfig'

// 用户下拉选项框类型
type UserSelect = {
  label: string
  value: number
```

```
}
const userList = ref<UserSelect[]>([])

const getUserList = async () => {
  let { data } = await UserApi.findAllUser()
  data.forEach((item: UserType) => {
    const info = {
      value: item.pkId,
      label: item.nickname
    }
    userList.value.push(info)
  })
}

// 初始化加载所有用户数据
onMounted(() => getUserList())

// 获取 ProTable 元素,调用其获取刷新数据方法(还能获取当前查询参数,方便导出携带参数)
const proTable = ref()

// 如果表格需要初始化请求参数,直接定义传给 ProTable(之后每次请求都会自动带上该参数,此参数
// 更改之后也会一直带上,改变此参数会自动刷新表格数据)
const initParam = reactive({})

// dataCallback 是对于返回的表格数据做处理,如果你后台返回的数据不是 list && total 这些字段,那么
// 你可以在这里处理成这些字段
const dataCallback = (data: any) => {
  return {
    list: data.list,
    total: data.total
  }
}

// 如果你想在请求之前对当前请求参数做一些操作,可以自定义如下函数:params 为当前所有的请求参
// 数(包括分页),最后返回请求列表接口
// 默认不做操作就直接在 ProTable 组件上绑定 :requestApi = "getUserList"
const getTableList = (params: any) => {
  let newParams = { ...params }
```

```
    return getResourcePage(newParams)
}


// 表格配置项
const columns: ColumnProps<SysResource.ResResourceList>[] = [
    { type: 'selection', fixed: 'left', width: 60 },
    {
        prop: 'title',
        label: '资源标题',
        search: { el: 'input' }
    },
    {
        prop: 'author',
        label: '作者',
        showOverflowTooltip: true,
        search: { el: 'select' },
        enum: userList.value,
        fieldNames: { label: 'label', value: 'value' }
    },
    {
        prop: 'diskType',
        label: '网盘分类',
        width: 200
    },
    {
        prop: 'resTypeList',
        label: '资源分类',
        showOverflowTooltip: true,
        width: 300,
        render: (scope) => {
            return scope.row.resTypeList.map((item: string) => {
                return (
                    <el-tag type="success" class="mr-2" effect="dark">
                        {item}
                    </el-tag>
                )
            })
        }
```

```
    },
    {
        prop: 'tagList',
        label: '标签',
        showOverflowTooltip: true,
        width: 300,
        render: (scope) => {
            let tagList = scope.row.tagList
            return (
                <div>
                    {tagList.map((item: string) => {
                        return (
                            <el-tag type="info" class="mr-2" effect="dark">
                                {item}
                            </el-tag>
                        )
                    })}
                </div>
            )
        }
    },
    {
        prop: 'isTop',
        label: '是否置顶',
        width: 100,
        search: { el: 'select', props: { filterable: true } },
        // enum: [
        //     { title: '是', value: 1 },
        //     { title: '否', value: 0 }
        // ],
        enum: () => dictConfigList('isTop'),
        fieldNames: { label: 'title', value: 'value' },
        render: (scope) => {
            return (
                <el-tag type={scope.row.isTop === 0 ? 'success' : 'warning'} effect round>
                    {scope.row.isTop === 1 ? '是' : '否'}
                </el-tag>
            )
        }
```

```
      },
    },
    {
      prop: 'price',
      label: '积分'
    },
    {
      prop: 'likeNum',
      label: '点赞数'
    },
    {
      prop: 'detail',
      showOverflowTooltip: true,
      label: '详情'
    },
    {
      prop: 'downloadUrl',
      showOverflowTooltip: true,
      label: '下载链接'
    },
    {
      prop: 'status',
      label: '审核状态',
      width: 150,
      search: { el: 'select', props: { filterable: true } },
      fieldNames: { label: 'title', value: 'value' },
      // enum: [
      //   { title: '待审核', value: 0 },
      //   { title: '审核通过', value: 1 },
      //   { title: '审核不通过', value: 2 }
      // ],
      enum: () => dictConfigList('status'),
      render: (scope) => {
        let type = scope.row.status === 0 ? 'warning' : scope.row.status === 1 ? 'success' : 'danger'
        return <el-tag type={type}>{scope.row.status === 0 ? '待审核' : scope.row.status === 1 ? '审核通过' : '审核不通过'}</el-tag>
      }
    },
```

```
        {
            width: 200,
            prop: 'remark',
            showOverflowTooltip: true,
            label: '审核结果描述'
        },
        {
            prop: 'createTime',
            label: '创建时间',
            width: 200
        },
        { prop: 'operation', label: '操作', fixed: 'right', width: 100 }
    ]
</script>
```

获取数据,页面效果如图1所示。

图1 搭建资源列表页面效果图

2 资源审核

2.1 任务描述

在开发资源审核功能时,波哥为小南和小工分配了不同的任务内容。

波哥负责后端的 DTO 请求实体封装,在 model.dto 包中新建一个 ResourceAuditDTO 类,用于封装审核所需的关键参数。这些参数将确保后端能够准确地接收前端传递的审核信息。

小南负责在 Service 层新增审核方法。在 ResourceService 接口中定义一个新的审核方法,并在 ResourceServiceImpl 类中实现这个方法。由于具体的审核逻辑尚未完成,小南将在实现中留下一个 TODO 标记,作为后续工作的提醒。

与此同时,小工开始前端的工作。在后台管理系统前端项目中搭建审核投稿组件。这个组件将是一个表单审核页面,允许用户选择是否审核通过,并输入对应的审核描述。

在本项目中,表单审核状态有一个自定义的验证规则。默认情况下,如果校验不符合要求,就需要进行自定义处理。小工将确保在审核表单中,当属性值不是 1 或 2 时,表单验证不通过,并给出相应的提示信息。小工还将修改资源弹框组件 src/views/Resource/components/ResourceDialog.vue,以便在需要时展示或隐藏审核相关的内容。

小工还要负责在父组件中调用审核投稿组件。当用户点击审核按钮时,父组件将传递当前资源的 ID 给审核组件,并调用审核组件进行审核操作。这样,用户就可以直接在父组件中完成资源的审核流程,无需跳转到其他页面。

2.2 任务分析

2.2.1 DTO 请求实体封装

在 model.dto 包中新建 ResourceAuditDTO,封装审核所需参数。

```
@Data
@Schema(name = "resourceAuditDTO", description = "资源审核 dto")
public class ResourceAuditDTO {
    @NotNull
    @Schema(name = "pkId", description = "主键")
    private Integer pkId;

    @Schema(name = "status", description = "状态")
    private Integer status;

    @Schema(name = "remark", description = "审核描述")
    private String remark;
}
```

2.2.2 Service 层新增审核方法

1. ResourceService 接口新增方法。

```
void audit(ResourceAuditDTO dto);
```

2. ResourceServiceImpl 实现新定义的方法,留下一个 TODO 标记,后面实现。

```java
@Override
public void audit(ResourceAuditDTO dto) {
    Resource resource = baseMapper.selectById(dto.getPkId());
    if (ObjectUtils.isEmpty(resource)) {
        log.error("资源不存在, pkId:{}", dto.getPkId());
        throw new ServerException("资源不存在");
    }
    resource.setStatus(dto.getStatus());
    resource.setRemark(dto.getRemark());
    baseMapper.updateById(resource);
    User author = userService.getById(resource.getAuthor());
    if (ObjectUtils.isEmpty(author)) {
        log.error("作者不存在, pkId:{}", resource.getAuthor());
        throw new ServerException("作者不存在");
    }

    // TODO 新增用户积分
}
```

2.2.3 Controller 层接口实现

ResourceController 新增审核接口。

```java
@PostMapping("/audit")
@Operation(summary = "审核")
@PreAuthorize("hasAuthority('sys:resource:audit')")
public Result<Void> audit(@RequestBody @Valid ResourceAuditDTO dto) {
    resourceService.audit(dto);
    return Result.ok();
}
```

2.2.4 前端搭建审核投稿组件

审核投稿页面是一个表单审核,选择是否审核通过,并输入对应的审核描述,具体属性可以查看官网。

本项目的表单审核状态进行了自定义验证规则,默认情况下,校验不符合要求,就需要自定义处理,判断属性值是否为 1 或 2,如果是其他就不通过验证。修改资源弹框组件 src/

views/Resource/components/ResourceDialog.vue。

```vue
<template>
    <Dialog :model-value="dialogVisible" :title="dialogProps.title" :fullscreen="dialogProps.fullscreen" :
max-height="dialogProps.maxHeight" :cancel-dialog="cancelDialog">
        <div :style="'width: calc(100% - ' + dialogProps.labelWidth! / 2 + 'px)'">
            <el-form
                ref="ruleFormRef"
                label-position="right"
                :label-width="dialogProps.labelWidth + 'px'"
                :rules="rules"
                :model="dialogProps.row"
                :disabled="dialogProps.isView"
                :hide-required-asterisk="dialogProps.isView"
            >
                <el-form-item label="审核状态" prop="status">
                    <el-radio-group v-model="dialogProps.row!.status">
                        <el-radio :label="1" border>审核通过</el-radio>
                        <el-radio :label="2" border>审核不通过</el-radio>
                    </el-radio-group>
                </el-form-item>
                <el-form-item label="审核描述" prop="remark">
                    <el-input v-model="dialogProps.row.remark" type="textarea" :rows="3" :disabled="dialogProps.isView" placeholder="请输入审核意见长度需要小于 200 字" />
                </el-form-item>
            </el-form>
        </div>
        <template #footer>
            <slot name="footer">
                <el-button @click="cancelDialog">取消</el-button>
                <el-button type="primary" v-show="!dialogProps.isView" @click="handleSubmit">确定</el-button>
            </slot>
        </template>
    </Dialog>
</template>

<script setup lang="ts">
import { ref, reactive } from 'vue'
```

```
import { ElMessage, FormInstance } from 'element-plus'
import { Dialog } from '@/components/Dialog'
interface DialogProps {
    title: string
    isView: boolean
    fullscreen?: boolean
    row: any
    labelWidth?: number
    maxHeight?: number | string
    api?: (params: any) => Promise<any>
    getTableList?: () => Promise<any>
    roleList?: any
}
const dialogVisible = ref(false)
const dialogProps = ref<DialogProps>({
    isView: false,
    title: '',
    row: { status: 1, remark: '' },
    labelWidth: 160,
    fullscreen: true,
    maxHeight: '500px'
})


// 接收父组件传过来的参数
const acceptParams = (params: DialogProps): void => {
    dialogProps.value.row = { ...dialogProps.value.row, ...params.row }
    dialogProps.value = { ...dialogProps.value, ...params }
    dialogVisible.value = true
}


defineExpose({
    acceptParams
})


const rules = reactive({
    status: [
        { required: true, message: '请选择是否通过审核', trigger: 'blur' },
        {
```

```
      validator: (rule: any, value: any, callback: any) => {
        if (value === 1 || value === 2) {
          callback()
        } else {
          callback(new Error('请选择是否通过审核'))
        }
      },
      trigger: 'blur'
    }
  ],
  remark: [
    { required: true, message: '请输入审核描述', trigger: 'blur' },
    {
      min: 3,
      max: 200,
      message: '长度在 3 到 200 个字符',
      trigger: 'blur'
    }
  ]
})
const ruleFormRef = ref<FormInstance>()
const handleSubmit = () => {
  ruleFormRef.value!.validate(async (valid) => {
    if (!valid) return
    try {
      await dialogProps.value.api!(dialogProps.value.row).then((_) => {
        ElMessage.success({ message: `${dialogProps.value.title}成功!` })
        dialogProps.value.getTableList!()
        dialogVisible.value = false
        ruleFormRef.value!.resetFields()
      })
    } catch (error) {
      console.log(error)
    }
  })
}
const cancelDialog = () => {
  dialogVisible.value = false
```

```
        ruleFormRef.value!.resetFields()
   }
</script>

<style scoped lang="less"></style>
```

2.2.5 父组件调用审批投稿组件

直接在父组件点击审批按钮调用审批组件进行审批,传递当前ID。

```
<template>
    <ProTable ref="proTable" title="资源列表" :columns="columns" :requestApi="getTableList" :initParam="initParam" :dataCallback="dataCallback">
        <!-- 表格操作 -->
        <template #operation="scope">
            <el-button type="primary" link :icon="EditPen" v-hasPermi="['sys:resource:audit']" @click="openDrawer('审批', scope.row.pkId)" v-if="scope.row.status == 0">审批</el-button>
        </template>
    </ProTable>
    <ResourceDialog ref="dialogRef" />
</template>

<script setup lang="tsx" name="SysResource">
import ResourceDialog from './components/ResourceDialog.vue'
import { EditPen } from '@element-plus/icons-vue'
import { auditResource } from '@/api/modules/resource'

// 打开 drawer(新增、查看、编辑)
const dialogRef = ref()
const openDrawer = (title: string, pkId: number) => {
    let params = {
        title,
        row: { pkId },
        api: auditResource,
        getTableList: proTable.value.getTableList,
        maxHeight: '150px'
    }
    dialogRef.value.acceptParams(params)
}
</script>
```

点击"审批",页面效果如图 2 所示。审批完成后的页面效果如图 3 所示。

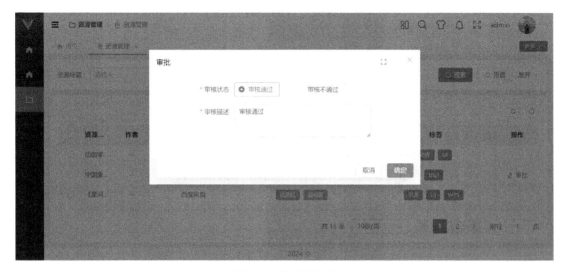

图 2　审批页面效果

图 3　审批成功页面效果

3　补充积分业务

3.1　任务描述

在之前的任务中,波哥已经带领小南和小工完成了积分管理模块的开发。随着业务的展开,他们需要对积分模块进行功能上的拓展。

波哥负责在 Service 层实现积分日志相关的业务逻辑。在 BonusLogService 接口中创建一个新增积分日志的方法,这个方法将是一个重载方法,拥有相同的方法名和返回值,但接受不同的参数。这将使得调用者可以根据需要选择是否传递积分值作为参数。接下来,在 BonusLogServiceImpl 类中实现这个接口方法。为了实现代码的复用和简化,需要创建一个名为 addBonusLogBase 的基础方法,用于处理数据库层的操作。两个重载的接口方法都将调用这个基础方法。在添加积分日志时,如果调用者没有传递积分值,波哥将使用枚举类中定义的价格作为默认值;如果传递了积分值,那么将使用传入的值进行增减操作。

小南负责在后台管理系统中补充积分业务的相关逻辑。在 ResourceServiceImpl 类的 audit 方法中的 TODO 标记处添加新增用户积分的代码。这个代码将根据审核结果(通过或拒绝)为用户增加或减少积分。小南将确保这段代码与波哥实现的积分日志业务逻辑相协调,确保在审核过程中能够正确记录积分的变化,并将积分更新到用户的账户中。

小工负责在前端展示和管理这些积分业务相关的功能。根据后端提供的接口和数据,搭建和配置相应的前端页面和组件,以便用户能够直观地查看和管理自己的积分。小工将确保前端与后端的交互顺畅,用户能够正常地进行积分的查看、获取和消费等操作。

3.2 任务分析

3.2.1 Service 层业务实现

1. BonusLogService 创建新增积分日志的方法,这里是一个重载,相同方法名、返回值、不同的方法入参。

```
void addBonusLog( Integer userId, BonusActionEnum contentEnum, Integer bonus);

void addBonusLog( Integer userId, BonusActionEnum contentEnum);
```

2. BonusLogServiceImpl 类,实现接口方法。

创建基础方法 addBonusLogBase 用于数据库层的操作,两个接口方法都调用基础方法。如果不传积分,那就直接用枚举类中定义的价格,否则使用入参进行增减。

```
private final UserService userService;

@Override
public void addBonusLog( Integer userId, BonusActionEnum contentEnum, Integer bonus) {
    addBonusLogBase( userId, contentEnum.getDesc( ), bonus);
}

@Override
public void addBonusLog( Integer userId, BonusActionEnum contentEnum) {
    addBonusLogBase( userId, contentEnum.getDesc( ), contentEnum.getBonus( ));
```

```java
    }

    private void addBonusLogBase(Integer userId, String content, Integer bonus) {
        BonusLog bonusLog = new BonusLog();
        bonusLog.setUserId(userId);
        bonusLog.setContent(content);
        bonusLog.setBonus(bonus);
        save(bonusLog);
        User user = userService.getById(userId);
        // 添加积分
        user.setBonus(user.getBonus() + bonus);
        userService.updateById(user);
    }
```

3.2.2 管理后台补充 TODO 新增积分

找出 ResourceServiceImpl 类中 audit 方法,在 TODO 标记处添加新增用户积分的代码。

```java
private final BonusLogService bonusLogService;


@Override
@Transactional(rollbackFor = Exception.class)
public void audit(ResourceAuditDTO dto) {
    Resource resource = baseMapper.selectById(dto.getPkId());
    if (ObjectUtils.isEmpty(resource)) {
        log.error("资源不存在, pkId:{}", dto.getPkId());
        throw new ServerException("资源不存在");
    }
    resource.setStatus(dto.getStatus());
    resource.setRemark(dto.getRemark());
    baseMapper.updateById(resource);
    User author = userService.getById(resource.getAuthor());
    if (ObjectUtils.isEmpty(author)) {
        log.error("作者不存在, pkId:{}", resource.getAuthor());
        throw new ServerException("作者不存在");
    }


    // 新增用户积分
    bonusLogService.addBonusLog(author.getPkId(), BonusActionEnum.RESOURCE_AUDIT_PASS);

}
```

4 任务总结

在本次的开发任务中,我们成功完成了资源列表、资源审核以及补充积分业务三个小节的开发工作。

一、资源列表

在资源列表的开发过程中,我们优化了后台管理系统的资源展示功能。我们通过合理的数据结构和前端展示逻辑,实现了资源的清晰分类和高效检索。用户能够直观地查看各类资源信息,并进行快速检索。这不仅提升了用户体验,也提高了工作效率。

二、资源审核

在资源审核功能的实现上,我们注重了流程的完整性和逻辑的严谨性。我们通过在后端 Service 层定义审核相关的业务逻辑,并在 Controller 层提供接口供前端调用,实现了资源审核的自动化和标准化。同时,我们在前端搭建了审核投稿组件,并设置了自定义验证规则,确保用户提交的审核信息准确无误。这一功能的实现,不仅提升了资源的质量,也确保了平台的稳定和安全。

三、补充积分业务

在补充积分业务的开发中,我们重点解决了积分的增减和记录问题。我们通过在 Service 层定义新增积分日志的方法,并在实现类中调用基础方法处理数据库操作,确保了积分变化的实时性和准确性。同时,我们在后台管理系统中补充了新增用户积分的代码,使得用户在通过资源审核后能够自动获得相应的积分奖励。这一功能的实现,不仅丰富了平台的激励机制,也提升了用户的活跃度和黏性。

任务十

首页仪表盘开发

波哥向小南和小工展示一个空白的首页大屏界面:"看,这是我们管理后台的首页,我们需要加些图表来展示数据。"

小南:"嗯,我知道,比如标签词云图、今日签到、资源总数、资源占比图饼图展示等。"

小工:"听起来不错,那我们开始吧!"

◇ 任务点

- 首页仪表盘数据构造;
- 前端图表运用。

◇ 任务计划

- 任务内容:首页仪表盘数据展示;
- 任务耗时:预计完成时间为 1~2 h;
- 任务难点:前端图表的使用。

1 仪表盘效果展示

1.1 任务描述

波哥对小南和小工说:"我初步设计了一个首页仪表盘的内容,如果大家没有其他想法的话,我们就先按照这个模板进行设计和开发。等后面有了新的思路,可以继续扩展我们的首页仪表盘的内容。"

1.2 任务分析

首页仪表盘分为三个部分:左边是标签词云图,中间部分是一些数值展示,右边部分是不同网盘的资源占比图(图1)。仪表盘使用组件为 Echarts,如需查看对应属性及图表可以搜索 Echarts 官网。

图 1　首页仪表盘的内容

2　首页仪表盘后端实现

2.1　任务描述

在首页仪表盘后端实现工作中,波哥、小南和小工将按照以下步骤协作完成。

波哥首先明确了仪表盘后端所需的数据结构和功能点,并与前端团队协商确定了 DashboardDataVO 的数据视图结构。

小南负责数据层的工作。对于标签词云图的数据准备,他认识到这部分数据较为简单,可以直接利用 Mybatis Plus 在 IService 接口中封装的 list 方法获取,无需额外处理。同样,用户总数和资源总数的获取也通过框架封装的 count 方法实现,高效且直接。

小工承担业务逻辑层和接口实现的任务。在数值展示区的"今日签到数"部分,首先在 BonusLogService 接口中新增了获取今日签到用户数量的方法,并在 BonusLogServiceImpl 中实现了该方法,通过查询积分日志表来获取结果。对于"未审批资源总数",小工在 ResourceService 接口中新增了按状态查询数量的方法,并在 ResourceServiceImpl 中实现,确保能够精确获取未审批的资源数量。

在资源占比图的数据准备上,小工在 ResourceMapper 接口中新增了按网盘类型获取数据的方法,并在 ResourceMapper.xml 文件中实现了具体的 SQL 查询语句。在 ResourceService 接口中新增了获取不同网盘资源数量的方法,并在 ResourceServiceImpl 中实现,确保能够按需获取不同网盘类型的资源数据。

在 Service 层,波哥指导小工新建了 IndexService 接口,并添加了获取数据大屏所需数据的方法。在 IndexServiceImpl 实现类中实现了该方法,将各个部分的数据进行整合,形成完整的仪表盘数据。

在 Controller 层,小工在 IndexController 接口中新增了数据大屏数据接口,用于向前端提供仪表盘所需的数据。

2.2 任务分析

2.2.1 VO 返回视图封装

在 vo 包中创建返回实体 DashboardDataVO,定义和前端商议后得出的数据结构。如果想要展示更多,可以自己尝试新增属性。

```java
@Data
@Schema(name = "DashboardDataVO", description = "首页数据vo")
public class DashboardDataVO {

    @Schema(name = "userCount", description = "标签列表")
    private List<Tag> tagList;
    @Schema(name = "allUserCount", description = "用户总数")
    private Long allUserCount;
    @Schema(name = "todayUserCheckCount", description = "今日签到数")
    private Long todayUserCheckCount;
    @Schema(name = "resourceCount", description = "资源总数")
    private Long resourceCount;
    @Schema(name = "resourceUnAuditCount", description = "未审批资源总数")
    private Long resourceUnAuditCount;
    @Schema(name = "resourceCount", description = "不同网盘类型资源数量")
    private Map<String, Long> resourceCountMap;

}
```

2.2.2 标签词云图数据准备

标签数据比较简单,直接通过 Mybatis Plus 在 IService 中封装的 list 方法,这边就不对返回结果进行处理了,所以这块不需要写代码。

2.2.3 数值展示区的用户总数

和标签数据类似,直接使用封装好的 count 方法获取用户总数。

2.2.4 数据展示区的今日签到数

1. BonusLogService 新增方法,获取今日签到的用户。

```java
Long todayUserCheckCount();
```

2. BonusLogServiceImpl 实现方法，查询积分日志中全部的今日签到的数量。

```
@Override
public Long todayUserCheckCount() {
    Date now = new Date();

    // 设置开始时间和结束时间
    Date start = DateUtils.truncate(now, Calendar.DATE);
    Date end = DateUtils.addMilliseconds(DateUtils.ceiling(now, Calendar.DATE), -1);

    // 构建查询条件并执行查询
    return count(new LambdaQueryWrapper<BonusLog>()
            .eq(BonusLog::getContent, BonusActionEnum.DAILY_SIGN.getDesc())
            .between(BonusLog::getCreateTime, start, end));
}
```

2.2.5 数值展示区的资源总数

资源总数的实现方案和用户总数相同，可以直接使用框架封装好的方法。

2.2.6 数据展示区的未审批资源总数

1. ResourceService 新增根据状态查询数量的方法。

```
Long countResourceByStatus(ResourceStatusEnum status);
```

2. ResourceServiceImpl 实现方法。

```
@Override
public Long countResourceByStatus(ResourceStatusEnum status) {
    return baseMapper.selectCount(new LambdaQueryWrapper<Resource>().eq(Resource::getStatus, status.getCode()));
}
```

2.2.7 资源占比图

1. ResourceMapper 接口新增查询方法。

```
List<Map<String, Object>> selectCountByDiskType();
```

2. resources/mapper 文件夹下新建 ResourceMapper.xml，实现 mapper 层方法。

```xml
<select id="selectCountByDiskType" resultType="java.util.Map">
    select count(1) as count, tc.title as diskType from t_resource tr
```

```
        left join t_category tc on tc.type = 0 and tc.pk_id = tr.disk_type
        where tr.delete_flag = 0
        group by tr.disk_type
</select>
```

3. ResourceService 新增获取不同网盘类型资源数量的方法。

```
Map<String, Long> countResourceByDisk();
```

4. ResourceServiceImpl 实现方法。

```
@Override
public Map<String, Long> countResourceByDisk() {
    List<Map<String, Object>> list = baseMapper.selectCountByDiskType();
    if (list.isEmpty()) {
        return Collections.emptyMap();
    }
    return list.stream()
        .collect(Collectors.toMap(
            item -> item.get("diskType").toString(),
            item -> Long.parseLong(item.get("count").toString())
        ));
}
```

2.2.8 Service 层数据整合业务

1. 在 service 包中新建 IndexService 接口，新建获取数据大屏数据方法。

```
public interface IndexService {

    DashboardDataVO getDashboardData();

}
```

2. 在 service.impl 包中新建 IndexServiceImpl 实现类，实现接口定义的方法。

```
@Slf4j
@Service
@AllArgsConstructor
public class IndexServiceImpl implements IndexService {
    private final UserService userService;
    private final TagService tagService;
    private final ResourceService resourceService;
```

```java
    private final BonusLogService bonusLogService;

    @Override
    public DashboardDataVO getDashboardData() {
        DashboardDataVO resultVo = new DashboardDataVO();
        // 获取标签列表
        resultVo.setTagList(tagService.list());
        // 获取用户总数
        resultVo.setAllUserCount(userService.count());
        // 获取今日签到数
        resultVo.setTodayUserCheckCount(bonusLogService.todayUserCheckCount());
        // 获取资源总数
        resultVo.setResourceCount(resourceService.count());
        // 未审批资源总数
        resultVo.setResourceUnAuditCount(resourceService.countResourceByStatus(ResourceStatusEnum.UNAUDITED));
        // 获取资源 map
        resultVo.setResourceCountMap(resourceService.countResourceByDisk());
        return resultVo;
    }
}
```

2.2.9 Controller 层接口实现

IndexController 接口新增数据大屏数据接口。

```java
@RestController
@RequestMapping("index")
@Tag(name = "首页", description = "首页信息")
@AllArgsConstructor
public class IndexController {
    private final IndexService indexService;

    @GetMapping()
    @Operation(summary = "欢迎")
    public String index() {
        return "您好,项目已启动,祝您使用愉快!";
    }
}
```

```
@GetMapping("dashboard")
@Operation(summary = "首页数据")
public Result<DashboardDataVO> dashboard() {
    return Result.ok(indexService.getDashboardData());
}
```

3 首页仪表盘前端实现

3.1 任务描述

在首页仪表盘前端实现工作中,波哥、小南和小工共同协作,确保仪表盘页面的顺利搭建和数据展示。

波哥首先明确了仪表盘页面的整体布局和样式要求,并指出需要对现有页面 src/views/Home/Home.vue 进行修改。

小南负责样式和组件的编写工作。抽离了原有的样式代码,新建了 src/views/Home/index.less 样式文件,并在其中编写了仪表盘页面的样式。根据设计图,新建标签词云图组件 src/views/Home/components/wordCloud.vue,并在其中放置了静态数据以查看在页面中的初步效果。波哥审核了样式和组件的初步实现,并提出了改进意见。

小工则负责接口定义和数据获取的工作。新建了首页接口文件 src/api/modules/home/index.ts,并在其中提前定义了首页数据大屏所需的数据类型和相关接口。随后编写数据转换逻辑,将后端返回的数据(如标签和网盘数据)转化为前端所需的格式(如名称和值),并处理了 Map 结构的数据。

在数据获取和转换完成后,小工接着渲染中间数据区域,将获取的数据替换为真实数据并展示在页面上。同时,他也负责将父组件中的数据传递给标签词云图组件。标签词云图组件定义了接收数据的类型和监听函数,当数据发生改变时,会调用函数进行重新渲染。

最后,小工根据需求新建了饼图组件 src/views/Home/components/pie.vue,并查阅了 ECharts 官网了解饼图的详细属性和配置方法。他编写了饼图的渲染逻辑,并在父组件中将数据传递给饼图组件进行展示。波哥、小南和小工共同查看了页面的最终效果,并根据需要进行了微调。

3.2 任务分析

3.2.1 页面布局

搭建仪表盘前,我们需要修改之前页面,页面效果为之前效果图,修改首页 src/views/

Home/Home.vue。

```
<template>
  <div class="dataVisualize-box">
    <div class="card top-box">
      <div class="top-content">
        <el-row :gutter="40">
          <el-col class="mb40" :xs="24" :sm="12" :md="12" :lg="6" :xl="6">
            <div class="item-left sle">
              词云图
            </div>
          </el-col>
          <el-col class="mb40" :xs="24" :sm="12" :md="12" :lg="8" :xl="8">
            <div class="item-center">
              <div class="all-user traffic-box">
                <div class="traffic-img">
                  <el-icon><User /></el-icon>
                </div>
                <span class="item-value">99</span>
                <span class="traffic-name sle">用户总数</span>
              </div>
              <div class="no-resource traffic-box">
                <div class="traffic-img">
                  <el-icon><Warning /></el-icon>
                </div>
                <span class="item-value">99</span>
                <span class="traffic-name sle">未审批资源总数</span>
              </div>
              <div class="today-traffic traffic-box">
                <div class="traffic-img">
                  <el-icon><Location /></el-icon>
                </div>
                <span class="item-value">99</span>
                <span class="traffic-name sle">今日签到数</span>
              </div>
              <div class="all-resource traffic-box">
                <div class="traffic-img">
                  <el-icon><Files /></el-icon>
                </div>
```

```
                <span class="item-value">99</span>
                <span class="traffic-name sle">资源总数</span>
              </div>
            </div>
          </el-col>
          <el-col class="mb40" :xs="24" :sm="24" :md="24" :lg="10" :xl="10">
            <div class="item-right">
              <div class="echarts-title">不同网盘类型资源数量</div>
              <div class="pie-echarts">饼图</div>
            </div>
          </el-col>
        </el-row>
      </div>
    </div>
  </div>
</template>
<script setup lang="ts" name="dataVisualize">
import { User, Location, Files, Warning } from '@element-plus/icons-vue'
</script>


<style scoped lang="less">
@import url('./index.less');
</style>
```

首页代码较多,我们可以抽离样式代码,新建首页样式 src/views/Home/index.less 编写后,再引入到首页样式部分。

```
.dataVisualize-box {
  .top-box {
    box-sizing: border-box;
    padding: 25px 40px 0;
    margin-bottom: 10px;
    .top-title {
      margin-bottom: 10px;
      font-family: DIN;
      font-size: 18px;
      font-weight: bold;
    }
    .top-content {
```

```css
    margin-top: 10px;
  .el-icon {
    font-size: 30px;
  }
  .item-left {
    box-sizing: border-box;
    height: 100%;
    padding: 40px 0 30px 30px;
    overflow: hidden;
    color: #000;
    background-position: 50%;
    background-size: cover;
    border-radius: 20px;
    .left-title {
      font-family: DIN;
      font-size: 20px;
    }
    .img-box {
      display: flex;
      align-items: center;
      justify-content: center;
      width: 90px;
      height: 90px;
      margin: 40px 0 20px;
      background-color: #ffffff;
      border-radius: 20px;
      box-shadow: 0 10px 20px rgb(0 0 0 / 14%);
    }
    .left-number {
      overflow: hidden;
      font-family: DIN;
      font-size: 62px;
    }
  }
  .item-center {
    display: flex;
    flex-wrap: wrap;
    place-content: space-between space-between;
```

```css
    height: 100%;
  .traffic-box {
    box-sizing: border-box;
    display: flex;
    flex-direction: column;
    width: 47%;
    height: 48%;
    padding: 25px;
    border-radius: 30px;
    .traffic-img {
      display: flex;
      align-items: center;
      justify-content: center;
      width: 70px;
      height: 70px;
      margin-bottom: 10px;
      background-color: #ffffff;
      border-radius: 19px;
    }
  }
  .item-value {
    margin-bottom: 4px;
    font-family: DIN;
    font-size: 28px;
    font-weight: bold;
    color: #1a1a37;
  }
  .traffic-name {
    overflow: hidden;
    font-family: DIN;
    font-size: 15px;
    color: #1a1a37;
    white-space: nowrap;
  }
  .all-user {
    background-color: #e8faea;
    background-size: 100% 100%;
    .el-icon {
```

```css
      color: #89ef95;
    }
  }
  .no-resource {
    background-color: #e7e1fb;
    .el-icon {
      color: #a294d0;
    }
  }
  .today-traffic {
    background-color: #fdf3e9;
    .el-icon {
      color: #c4ad97;
    }
  }
  .all-resource {
    background-color: #f0f5fb;
    .el-icon {
      color: #d5e0ed;
    }
  }
}
.item-right {
  box-sizing: border-box;
  display: flex;
  flex-direction: column;
  width: 100%;
  height: 430px;
  border: 1px solid var(--el-border-color);
  border-radius: 25px;
  .echarts-title {
    padding: 15px 20px;
    font-family: DIN;
    font-size: 18px;
    font-weight: bold;
    border-bottom: 1px solid var(--el-border-color);
  }
  .pie-echarts {
```

```
          flex: 1;
          width: 100%;
        }
      }
    }
  }
  .bottom-box {
    position: relative;
    padding: 20px 0 0;
    .bottom-title {
      position: absolute;
      top: 75px;
      left: 50px;
      font-family: DIN;
      font-size: 18px;
      font-weight: bold;
    }
    .bottom-tabs {
      padding: 0 50px;
    }
    .curve-echarts {
      box-sizing: border-box;
      height: 400px;
      padding: 0 50px;
    }
  }
}
```

3.2.2 标签词云图

实现标签词云图,首先需要安装依赖。

```
npm i echarts echarts-wordcloud
```

新建标签词云图组件 src/views/Home/components/wordCloud.vue,放置一些静态数据,查看页面中的效果。

```
<script setup lang="ts">
import * as echarts from 'echarts/core'
import 'echarts-wordcloud'
```

```
import { onMounted } from 'vue'

onMounted(() => {
    init()
})
const data = [
    { value: 67, name: '红腹角雉' },
    { value: 98, name: '麝牛' },
    { value: 97, name: '山舌鱼' },
    { value: 100, name: '羚羊' },
    { value: 37, name: '非洲王子' },
    { value: 83, name: '麋鹿' },
    { value: 60, name: '中华鲟' },
    { value: 42, name: '鲔鱼' },
    { value: 96, name: '射水鱼' },
    { value: 54, name: '果子狸' },
    { value: 33, name: '小春鱼' },
    { value: 84, name: '水獭' },
    { value: 86, name: '刺猬' }
]

// 初始化示例设置数据
const init = () => {
    const dom = document.querySelector('.wordCloud') as HTMLElement
    var myChart = echarts.init(dom)
    const option = {
        series: [
            {
                type: 'wordCloud',

                // 要绘制云的形状,默认是 circle,可选的参数有 cardioid、diamond、triangle-forward、triangle、
                // star
                shape: 'circle',

                // 保持 maskImage 的纵横比或 1∶1 的形状
                // 从 echarts-wordcloud@ 2.1.0 开始支持该选项
                keepAspect: false,
```

```
// 左/上/宽/高/右/下用于词云的定位
// 默认放置在中心,大小为 75% × 80%
left: 'center',
top: 'center',
width: '100%',
height: '100%',
right: null,
bottom: null,

// 数据中的值将映射文本的大小范围
// 默认值为最小 12px,最大 60px
sizeRange: [12, 60],

// 文字旋转范围和步进程度。文本将通过 rotationStep 45 在[-90,90]范围内随机旋转
rotationRange: [-90, 90],
rotationStep: 45,

// 网格大小(以像素为单位),用于标记画布的可用性
// 网格大小越大,单词之间的间隔就越大
gridSize: 8,

// 设置为 true 允许文字部分地绘制在画布之外
// 允许画比画布大的字
// 从 echarts-wordcloud@ 2.1.0 开始支持该选项
drawOutOfBound: false,

// 如果字体太大,无法显示文本,是否缩小文本。如果设置为 false,则文本将不被渲染。如果
// 设置为 true,文本将被缩小
shrinkToFit: false,

// 是否执行布局动画
// 当单词较多时禁用会导致 UI 阻塞
layoutAnimation: true,

// 全局文本样式
textStyle: {
    fontFamily: 'sans-serif',
    fontWeight: 'bold',
```

```
            // Color 可以是回调函数或颜色字符串
            color: function () {
              // 任意颜色
              return 'rgb(' + [Math.round(Math.random() * 160), Math.round(Math.random() * 160), Math.round(Math.random() * 160)].join(',') + ')'
            }
          },
          emphasis: {
            focus: 'self',
            textStyle: {
              textShadowBlur: 10,
              textShadowColor: '#333'
            }
          },
          data: data
        }
      ]
    }
    myChart.setOption(option)
  }
</script>


<template>
  <div class="wordCloud"></div>
</template>

<style lang="less" scoped>
.wordCloud {
  height: 100%;
  width: 100%;
}
</style>
```

父组件导入标签词云图组件,查看页面整体效果(图2)。

```
<template>
<div class="item-left sle">
  <WordCloud />
</div>
```

```
</template>

<script setup lang="ts" name="dataVisualize">
import WordCloud from './components/wordCloud.vue'
</script>
```

图 2　导入标签词云图后的页面整体效果

3.2.3　定义类型声明和接口

新建首页接口 src/api/modules/home/index.ts,提前定义首页数据大屏相关类型和相关接口。

```
import http from '@/api'
import { _API } from '@/api/axios/servicePort'
import { SysTag } from '@/api/interface'
export type EChartsDataType = {
  name: string
  value: number
}

export type IndexDataReturnType = {
  allUserCount: number
  todayUserCheckCount: number
```

```
  resourceCount: number
  resourceUnAuditCount: number
  resourceCountMap: EChartsDataType
  tagList: SysTag.ResTagList[ ]
}
// 获取首页数据
export const indexDashboard = ( ) => {
  return http.get<IndexDataReturnType>(_API + '/index/dashboard')
}
```

3.2.4 首页获取数据

图3　结果返回类型对比图

根据接口的返回值,我们需要对返回的数据结构进行相应的转换。tagList 返回的是标签的相关信息,标签词云图只需要标签的标题及主键。根据主键,我们可以确定在标签词云图中各项的显示大小。resourceCountMap 返回的是每个网盘的资源数量,我们可以看见返回的是一个 Map 结果,我们需要提取所有的键值时,将其转化为数组,方便展示饼图。

```
<script setup lang="ts" name="dataVisualize">
import { indexDashboard } from '@/api/modules/home'
import type { IndexDataReturnType, EChartsDataType } from '@/api/modules/home'
import { onMounted, ref } from 'vue'

const indexData = ref<IndexDataReturnType>()
const keyWordList = ref<EChartsDataType[ ]>([ ])
const pieDataList = ref<EChartsDataType[ ]>([ ])
const getIndexData = async ( ) => {
```

```
  const res = await indexDashboard()
  indexData.value = res.data

  // 处理标签词云图
  keyWordList.value = res.data.tagList.map((item) => {
    return {
      name: item.title,
      value: item.pkId
    }
  })

  // 饼图
  pieDataList.value = Object.keys(res.data.resourceCountMap).map((item) => {
    const value = res.data.resourceCountMap[item]
    return {
      name: item,
      value
    }
  })
}

onMounted(() => getIndexData())
</script>
```

3.2.5 渲染中间数据区域

我们将获取的中间数据区域进行数据替换,渲染为真实数据。

```
<template>
  <span class="item-value">{{ indexData?.allUserCount }}</span>
  <span class="traffic-name sle">用户总数</span>
  <span class="item-value">{{ indexData?.resourceUnAuditCount }}</span>
  <span class="traffic-name sle">未审批资源总数</span>
  <span class="item-value">{{ indexData?.todayUserCheckCount }}</span>
  <span class="traffic-name sle">今日签到数</span>
  <span class="item-value">{{ indexData?.resourceCount }}</span>
  <span class="traffic-name sle">资源总数</span>
</template>
```

替换后的效果如图4所示。

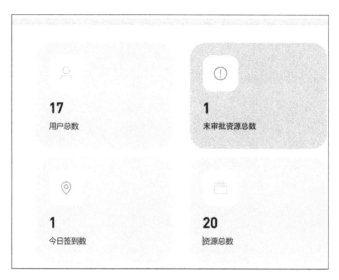

图 4　数据替换

3.2.6　父组件传值给标签词云图

标签词云图组件定义可接受的数据类型,以及监听传递的数据,当组件发生改变时,调用函数进行重新渲染。

```ts
<script setup lang="ts">
import { onMounted, watch } from 'vue'
import type { EChartsDataType } from '@/api/modules/home'

const props = defineProps<{
  keyWordList: EChartsDataType[]
}>()

onMounted(() => {
  init()
})

const init = () => {
  const option = {
    series: [
      {
        data: props.keyWordList
      }
    ]
  }
}
```

```
    myChart.setOption(option)
}

// 监听传递的数据
watch(
    () => props.keyWordList,
    () => {
        init()
    }
)
</script>
```

在父组件中,通过参数传递标签词云图数据。

```
<template>
    <WordCloud :keyWordList = "keyWordList" />
</template>
```

父组件传递获取之后的标签词云图数据的效果如图 5 所示。

图 5　标签词云图

3.2.7 饼图

饼图详细属性可以查看 Echarts 官网。

新建饼图组件 src/views/Home/components/pie.vue。

```
<script setup lang="ts" name="pie">
import type { EChartsDataType } from '@/api/modules/home'
import * as echarts from 'echarts'
import { onMounted, watch } from 'vue'
const props = defineProps<{
  pieDataList: EChartsDataType[]
}>()

onMounted(() => {
  init()
})
// 初始化
const init = () => {
  const dom = document.querySelector('.echarts') as HTMLElement
  var myChart = echarts.init(dom)
  const option = {
    // 标题相关
    title: {
      text: '各类网盘资源占比',
      subtext: '访问占比',
      left: '56%',
      top: '45%',
      textAlign: 'center',
      textStyle: {
        fontSize: 18,
        color: '#767676'
      },
      subtextStyle: {
        fontSize: 15,
        color: '#a1a1a1'
      }
    },
    tooltip: {
      trigger: 'item'
```

```
    },
    // 展示图例
    legend: {
      top: '4%',
      left: '2%',
      orient: 'vertical',
      icon: 'circle', //图例形状
      align: 'left',
      itemGap: 20,
      textStyle: {
        fontSize: 13,
        color: '#a1a1a1',
        fontWeight: 500
      },
      // 自定义不同演示的图例
      formatter: function (name: string) {
        let dataCopy = ''
        for (let i = 0; i < props.pieDataList.length; i++) {
          if (props.pieDataList[i].name == name && props.pieDataList[i].value >= 10000) {
            dataCopy = (props.pieDataList[i].value / 10000).toFixed(2)
            return name + '    ' + dataCopy + 'w'
          } else if (props.pieDataList[i].name == name) {
            dataCopy = props.pieDataList[i].value + ''
            return name + '    ' + dataCopy
          }
        }
        return ''
      }
    },
    // 展示数据
    series: [
      {
        type: 'pie',
        radius: ['70%', '40%'],
        center: ['57%', '52%'],
        silent: true,
        clockwise: true,
        startAngle: 150,
```

```
            data: props.pieDataList,
            labelLine: {
              length: 80,
              length2: 30,
              lineStyle: {
                width: 1
              }
            },
            // 标签相关
            label: {
              position: 'outside',
              show: true,
              formatter: '{d}%',
              fontWeight: 400,
              fontSize: 19,
              color: '#a1a1a1'
            },
            // 每一个显示风格颜色
            itemStyle: {
              normal: {
                color: function (params: any) {
                  //自定义颜色
                  var colorList = ['#feb791', '#fe8b4c', '#b898fd', '#8347fd', '#000', '#fff', '#feb791', '#fe8b4c', '#b898fd', '#8347fd', '#000', '#fff']
                  return colorList[params.dataIndex]
                }
              }
            }
          }
        ]
      }
      myChart.setOption(option)
    }

    // 监听数据变化
    watch(
      () => props.pieDataList,
      () => {
```

```
      init()
    }
  )
</script>
<template>
  <div class="echarts"> </div>
</template>

<style scoped>
.echarts {
  width: 100%;
  height: 100%;
}
</style>
```

父组件传值给子组件,查看页面效果。

```
<template>
<div class="pie-echarts">
  <Pie :pieDataList="pieDataList" />
</div>
</template>

<script setup lang="ts" name="dataVisualize">
import Pie from './components/pie.vue'
</script>
```

父组件传值给子组件的页面效果如图 6 所示。

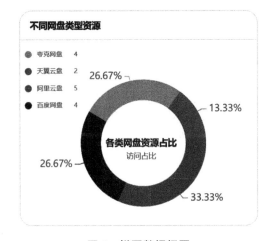

图 6 饼图数据视图

4 任务总结

在本次管理后台首页数据图表化展示的任务中，我们成功地将关键数据以直观、清晰的图表形式呈现在用户面前，大大提升了数据可读性和用户体验。整个任务分为仪表盘效果展示、后端实现和前端实现三个主要阶段，每个阶段都取得了显著的成果。

一、仪表盘效果展示

我们明确了仪表盘的整体布局和设计思路，确定了使用词云图展示标签数据，数值展示区展示今日签到数和资源总数等关键数据，以及饼图展示资源占比等信息。

二、首页仪表盘后端实现

我们实现了从数据库中提取数据、对数据进行清洗和转换，并最终以符合前端需求的格式返回给前端。

三、首页仪表盘前端实现

我们利用了前端框架和图表库的优势，快速高效地完成了页面的布局和组件的开发。我们抽离了样式代码，提高了代码的可维护性；通过定义类型声明和接口，确保了数据的准确性和一致性；实现了父子组件间的数据传递和组件的重用，提高了开发效率。最终，我们成功地将后端返回的数据以图表的形式展示在页面上，为用户提供了直观、清晰的数据视图。

任务十一

后台管理系统打包部署

波哥站在团队会议室的白板前,手持一支马克笔,说道:"现在,我们即将迎来一个关键的阶段——后台管理系统的打包部署。这不仅是技术层面的一次大考,更是对我们团队协作能力的检验。"

小南认真地看着白板上的内容,问:"波哥,这次任务我们具体需要做些什么呢?"

波哥开始在白板上画出流程图,说道:"首先,我们会一起搭建 Linux 环境,确保服务器配置正确、服务和库安装齐全。然后,开始后端项目的打包工作,确保代码和依赖项都整合到位。最后,进行前端项目的打包,确保网页、样式和脚本都能正常地显示和运行。我相信我们团队有能力顺利完成这次任务。在任务执行过程中,希望大家保持沟通和协作,确保每个环节都能顺利进行。如果有任何问题或困难,记得及时向我反馈。现在,我们开始行动吧!"

◇ 任务点

- 项目在 Linux 环境下的运行准备;
- 后端项目打包;
- 前端项目打包。

◇ 任务计划

- 开发内容:运行环境搭建、项目部署上线;
- 开发耗时:预计完成时间为 1~2 h;
- 开发难点:Linux 系统环境搭建。

1 后端环境搭建

1.1 任务描述

在后台管理系统打包部署工作中,波哥、小南和小工将共同协作,确保在 Linux 环境下顺利安装并配置必要的组件以支持后端服务的运行。

波哥首先明确了需要安装的三个关键组件:JDK 17、MySQL 8 和 Redis。他解释了这些组件的重要性:JDK 17 作为 Java 应用运行的基础,与后端服务项目的兼容至关重要;MySQL

8作为稳定的关系型数据库管理系统,用于存储和管理后台数据;而Redis则作为内存中的数据结构存储系统,用于提升系统性能和响应速度。

对于组件的安装顺序和安装过程中可能遇到的问题,波哥建议按照先JDK,后MySQL,再Redis的顺序进行安装,以确保基础服务的稳定性和后续组件的顺利运行。小南表示将查阅相关的安装文档和教程,确保安装过程顺利进行。

1.2 任务分析

1.2.1 JDK 17

在开发Java应用程序时,确保正确安装Java开发工具包(JDK)是非常重要的。以下是在Linux环境下安装JDK 17的详细步骤:

1. 下载安装包。从Oracle官网下载JDK 17的安装包。在终端使用wget命令从给定的URL下载。

```
wget https://download.oracle.com/java/17/latest/jdk-17_linux-x64_bin.tar.gz
```

这条命令将自动下载最新的JDK 17 Linux x64版本的安装包。

2. 解压安装包,修改包名为jdk-17。
- 使用tar命令解压下载的安装包:这将解压出一个包含JDK 17的文件夹。
- 清理安装包文件。
- 将解压出的文件夹重命名为jdk-17(注意:代码中的jdk-17.0.3.1是示例版本号,你需要根据实际的版本来重命名)。

```
tar zxf jdk-17_linux-x64_bin.tar.gz
rm -rf jdk-17_linux-x64_bin.tar.gz
mv jdk-17.0.3.1 jdk-17
```

3. 移动文件夹到/usr/local下。将jdk-17文件夹移动到/usr/local目录下,这是Linux系统的一个标准位置,用于存储本地安装的软件。

```
mv jdk-17 /usr/local/
```

4. 将Java添加到环境变量中。
- 使用vi编辑器(或其他你喜欢的文本编辑器)打开/etc/profile文件。
- 在文件的末尾添加以下两行内容来设置JAVA_HOME和PATH环境变量(注意:如果你的系统中已经安装了其他软件,并且这些软件也修改了PATH,请确保不要覆盖原有的内容)。
- 这里假设/usr/local/php/bin不在你的PATH中,如果它存在,并且你需要保留它,可以将其添加到PATH变量的定义中。

- 保存并关闭文件。

vi /etc/profile

文件添加的内容
export JAVA_HOME=/usr/local/jdk-17
export PATH=/usr/local/php/bin:/usr/local/jdk-17/bin:$PATH

5. 使配置生效。为了让刚才所做的环境变量更改立即生效,运行如下命令。你也可以简单地注销当前用户并重新登录。

source /etc/profile

6. 验证安装 JDK 版本。通过运行 java -version 命令来验证 JDK 是否已成功安装。

java -version

如果 JDK 安装成功,你将看到类似以下内容的输出,表明你已经成功安装了 JDK 17。

[root@iZbp1ik ~]# java -version
java version "17.0.9" 2023-10-17 LTS
Java(TM) SE Runtime Environment (build 17.0.9+11-LTS-201)
Java HotSpot(TM) 64-Bit Server (build 17.0.9+11-LTS-201, mixed mode, sharing)
[root@iZbp1ik ~]#

1.2.2 MySQL 8

1. 下载 MySQL 源包到服务器。我们需要在服务器上创建一个目录来存放 MySQL 的源,并使用 wget 命令下载 MySQL 8 的社区版源包(注意:URL 可能会随时间变化,请访问 MySQL 官网获取最新链接)。

mkdir mysql
cd mysql
wget https://repo.mysql.com/mysql80-community-release-el7-1.noarch.rpm

2. 安装下载的发行包。使用 rpm 命令安装下载的 MySQL 源包。安装完成后,可以通过 yum repolist enabled | grep "mysql.*-community.*" 来验证是否启用了 MySQL 的 yum 仓库。

rpm -Uvh mysql80-community-release-el7-1.noarch.rpm

3. 安装 MySQL。使用 yum 命令从 MySQL 的 yum 仓库中安装 MySQL 服务器。

yum -y install mysql-community-server --nogpgcheck

4. 启动 MySQL。安装完成后，使用 service 命令启动 MySQL 服务。

```
service mysqld start
```

5. 检查 MySQL 状态。使用 service 命令检查 MySQL 服务的状态。

```
service mysqld status
```

6. 查看 MySQL 密码。MySQL 8 在安装过程中会自动生成 root 用户的密码，并将该密码保存在日志文件中。

a. 新版 MySQL 安装之后会生成 root 用户的密码，该密码存储在 /var/log/mysqld.log 中

```
[root@ iZbp1ik ~]# grep 'temporary password' /var/log/mysqld.log
2024-10-02T08:17:16.645416Z6 [Note] [MY-010454] [Server] A temporary password is generated for root@localhost: Zqs#&nr12*)U
[root@ iZbp1ik ~]#
```

b. 使用 grep 命令查看密码

```
grep 'temporary password' /var/log/mysqld.log
```

7. 登录 MySQL 以及配置

- 登录和修改密码（登录时使用上一步复制的密码）。输入密码后，将登录到 MySQL 命令行界面。

```
mysql -u root -p
```

- 必须先修改密码，注意 MySQL 的密码策略已经改为中等，需要包括大小写字母、数字、特殊字符。MySQL 8 要求使用满足密码策略的强密码。这里暂时设置一个简单的密码（请注意，在实际环境中，应使用更复杂的密码）。

```
ALTER USER 'root'@'localhost' IDENTIFIED BY 'Abc.123!';
```

- 查看密码策略。根据策略要求，可能需要调整密码策略。

```
show variables like '%password%';
```

- 调整密码策略将密码策略的长度调整为 6 和密码的强度设置为低。如果默认的密码策略过于严格，可以临时降低它以允许使用较简单的密码。然后，就可以再次修改密码以满足新的策略要求。

```
set global validate_password.policy=LOW;
set global validate_password.length=6;
```

- 修改简单密码。

```
ALTER USER 'root'@'localhost' IDENTIFIED BY 'soft@123';
```

- 开启 MySQL 远程连接功能

a. 需要新建可远程的 root 用户,然后授权。创建一个可以从任何位置连接的新 root 用户(出于安全考虑,不建议在生产环境中这样做,但在测试和开发环境中可以)。

\# 这行命令是在 MySQL 数据库中创建一个名为 root 的用户,并指定这个用户可以从任何位置('%' 代表
\# 通配符,表示任何 IP 地址)连接到数据库,该用户的密码被设置为 soft@ 123
CREATE USER 'root'@'%' IDENTIFIED BY 'soft@ 123';

\# 这行命令是在 MySQL 数据库中授予名为 root 的用户在所有数据库和所有表上拥有全部权限
GRANT ALL PRIVILEGES ON *.* TO 'root'@'%';
FLUSH PRIVILEGES;

b. 调整 root@'%' 的身份认证方式。调整新创建的 root@'%' 用户的身份认证方式(MySQL 8 默认使用 caching_sha2_password,但某些客户端可能不支持,因此这里更改为 mysql_native_password)。

ALTER USER 'root'@'%' IDENTIFIED WITH mysql_native_password BY 'soft@ 123';

c. 测试连接。现在,应该可以从任何位置使用新的 root 用户和密码连接到 MySQL 服务器了。可以使用 MySQL 命令行工具或图形化工具(如 Navicat Premium)进行测试。如图 1 所示。

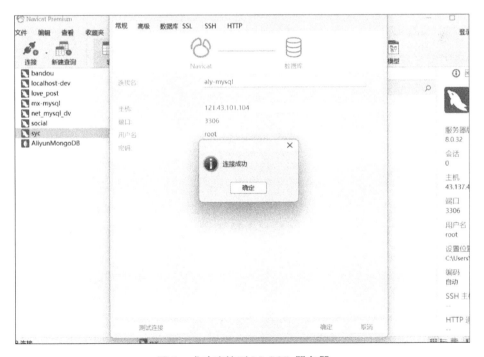

图 1　成功连接到 MySQL 服务器

1.2.3 Redis

1. 更新 YUM 源。在开始安装 Redis 之前,建议先更新系统的 YUM 源,以确保安装过程中使用的软件包是最新的。运行命令更新 YUM 源。

```
sudo yum update
```

如果在更新过程中遇到源的问题,可以尝试更换 YUM 源。以下是一个更换为阿里云镜像源的示例。

```
# 备份初始源配置
mv /etc/yum.repos.d/CentOS-Base.repo /etc/yum.repos.d/CentOS-Base.repo.backup

# 下载新的阿里云镜像源
curl -o /etc/yum.repos.d/CentOS-Base.repo http://mirrors.aliyun.com/repo/Centos-7.repo

# 清理和重建索引
yum clean all
yum makecache

# 重新回到第一步
sudo yum update
```

2. 安装 Redis 数据库。在 YUM 源更新完毕后,我们可以使用 YUM 包管理器来安装 Redis。这个命令会自动处理 Redis 相关的依赖关系,并将其安装到您的系统上。

```
sudo yum -y install redis
```

3. 启动 Redis 服务。安装完成后,我们需要启动 Redis 服务,运行以下命令启动 Redis 服务。

```
sudo systemctl start redis
```

4. 修改 Redis 配置,并设置连接密码。Redis 的默认配置文件位于 /etc/redis.conf。我们可以编辑这个文件来更改 Redis 的配置。使用 vi 或其他文本编辑器打开 Redis 配置文件。

```
vi /etc/redis.conf
```

a. 注释掉 bind 指令(如果存在),以便 Redis 可以接受来自任何 IP 地址的连接。找到类似 bind 127.0.0.1 的行,并在其前面添加 "#" 符号来注释它。

```
#bind 127.0.0.1
```

b. 取消注释 requirepass 指令，并为其设置一个密码。找到类似#requirepass foobared 的行，去掉前面的"#"符号，并将 foobared 替换为实际的密码。请将"密码"处替换为您选择的密码。

#requirepass foobared

requirepass 密码

5. 重启 Redis 服务。在更改了 Redis 的配置文件后，我们需要重启 Redis 服务以使更改生效。运行以下命令重启 Redis 服务：

sudo systemctl restart redis

现在，Redis 数据库已经安装并配置好密码保护。可以使用 Redis 客户端工具（如 rediscli）连接到 Redis 数据库，并使用之前设置的密码进行身份验证。

以下是 Redis 的一些常用操作命令：
- systemctl start redis.service #启动 redis 服务器；
- systemctl stop redis.service #停止 redis 服务器；
- systemctl restart redis.service #重新启动 redis 服务器；
- systemctl status redis.service #获取 redis 服务器的运行状态；
- systemctl enable redis.service #开机启动 redis 服务器；
- systemctl disable redis.service #开机禁用 redis 服务器。

2 后端 API 部署

2.1 任务描述

在后端 API 部署工作中，波哥、小南和小工将携手合作，确保后端 API 能够在服务器上顺利运行。

波哥强调了部署前的准备工作，包括修改项目配置文件中的特定环境设置，如数据库连接信息和服务器地址，以及使用构建工具对项目进行打包，生成可执行的 JAR 或 WAR 包。

小南根据波哥的指示，负责修改配置文件并进行项目打包。打包完成后，小南通过 FTP 或 SCP 协议将文件上传到服务器。一旦文件上传完毕，小南便立即通知小工进行后续部署工作。

小工接收到小南的通知后，首先检查上传的文件是否完整，并将其放置在服务器上的正确目录下。接着，小工配置服务器环境，包括设置端口号、防火墙规则等，以确保 API 能够正常访问和通信。完成环境配置后，小工运行小南上传的应用程序，并检查其是否能够正常运行，包括测试 API 接口的功能和性能。

2.2 任务分析

2.2.1 修改配置

在 application-dev. yml 文件中，有项目的一些开发环境的配置，需要改成线上地址，在环境搭建中我们搭建了线上 MySQL 和 Redis。

我们创建新的配置文件，拷贝 application-dev. yml 文件，重命名为 application-prod. yml，修改配置中的地址为线上地址(图 2)。

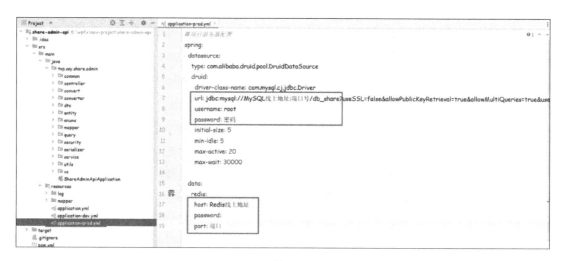

图 2　修改配置

2.2.2 项目打包

1. 在项目根目录下打开控制台，输入指令 mvn clean -DskipTests = true package，项目就会开始构建打包(图 3)。

```
G:\wpfx\share-admin-api
mvn clean -DskipTests=true package
[INFO] Scanning for projects...
[INFO]
[INFO] -----------------------< top.ssy:share-admin-api >-----------------------
[INFO] Building share-admin-api 1.0.0
[INFO]   from pom.xml
[INFO] --------------------------------[ jar ]---------------------------------
[INFO]
[INFO] --- clean:3.3.2:clean (default-clean) @ share-admin-api ---
[INFO] Deleting G:\wpfx\share-admin-api\target
```

图 3　输入项目打包指令

2. 运行成功后,在项目根目录会有一个 target 文件夹,文件夹中的 share-admin-api.jar 就是我们部署的文件(图4)。

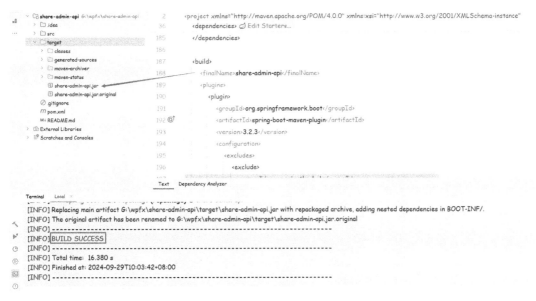

图4 得到 share-admin-api.jar 文件

在项目 pom.xml 文件中 build 节点下可以设置 finalName,就是打包后 JAR 包的名字。

2.2.3 文件上传

使用自己的 SSH/FTP 工具,把刚刚打包好的 JAR 包上传到服务器中。例如这里放在了 opt 文件夹下。

[root@ iZbp1ik ~]# find / -name share-admin-api.jar
/opt/share-admin-api.jar
[root@ iZbp1ik ~]#

2.2.4 运行

1. 进入 JAR 包所在的目录,使用指令 java -jar share-admin-api.jar --spring.profiles.active=prod 设置运行环境为 prod 开发环境。

[root@ iZbp1ik ~]# cd /opt
[root@ iZbp1ik opt]# java -jar share-admin-api.jar --spring.profiles.active=prod

运行成功后,可以在浏览器打开接口文档测试一下。但是这样的方式在服务器控制台关闭后项目就会停止,所以需要一个后台运行的方式,如 nohup java -jar /opt/share-admin-api.jar --spring.profiles.active=prod >/opt/app-log.file 2>&1 &。

```
[root@ iZbp1ik opt]# kill -9 851
[root@ iZbp1ik opt]# nohup java -jar /opt/share-admin-api.jar --spring.profiles.active=prod >/opt/app-log.file 2>&1 &
[root@ iZbp1ik opt]# lsof -i:8081
[root@ iZbp1ik opt]# lsof -i:8081
COMMAND PID USER FD TYPE DEVICE SIZE/OFF NODE NAME
java        851 root 14u IPV6 4867188   0t0      TCP *:tproxy (LISTEN)
[root@ iZbp1ik opt]# cat /opt/app-log.file
nohup: ignoring input
```

2. 使用 nohup 指令会让项目在后台运行,这个 app-log.file 就是项目的日志文件,运行成功后还是用接口文档测试一下。需要停止服务的话,使用 kill 指令,851 是服务运行的 PID,查询方法可以网上自己搜索。

```
[root@ iZbp1ik opt]# kill -9 851
[root@ iZbp1ik opt]# lsof -i:8081
[1]+ Killed   nohup java -jar /opt/share-admin-api.jar --spring.profiles.active=dev >/opt/app-log.file 2>&1
[root@ iZbp1ik opt]# lsof -i:8081
[root@ iZbp1ik opt]#
```

部署成功后就可以把服务地址给前端,记得需要关闭服务器防火墙,开放对应端口。

3 前端部署

3.1 任务描述

在前端部署工作中,波哥、小南和小工共同协作以确保前端应用能够成功运行在服务器上。

波哥向小南指示了前端部署的首要步骤,即确保服务器上安装了 Nginx,作为高性能的 HTTP 和反向代理服务器来承载前端应用。

小南根据波哥的指导,使用服务器上的包管理器(如 apt、yum 等)来安装 Nginx。安装完成后,波哥进一步指示小南配置 Nginx 的线上接口,包括编辑 Nginx 的配置文件,指定前端项目的根目录、端口号等关键信息。

配置 Nginx 后,波哥提醒小南进行前端项目的打包工作。前端项目通常使用构建工具(如 Webpack、Vue-CLI、Create React App 等)来生成静态文件。小南按照项目的要求使用相应的构建工具对前端项目进行打包,生成 HTML、CSS、JavaScript 等静态文件。

一旦前端项目打包完成,小南需要将生成的静态文件复制到 Nginx 指定的根目录下。

这样,当 Nginx 接收到来自客户端的请求时,它会从该目录下提供静态文件作为响应。

3.2 任务分析

3.2.1 安装 Nginx

在部署前端项目时,确保服务器正确安装 Nginx 是非常重要的,以下是在 Linux 环境下安装 Nginx 的详细步骤:

1. 安装依赖项。在开始安装 Nginx 之前,首先需要安装一些依赖项,以确保 Nginx 编译和运行正常。打开终端并执行以下命令:

```
yum install -y wget gcc-c++ pcre-devel zlib-devel openssl-devel
```

这将安装必要的工具和库,以支持 Nginx 的编译和运行。

2. 下载 Nginx。可以在 Nginx 官网上找到最新版本的下载链接(图 5)。

```
# 例如,下载 Nginx 1.24.0 版本
wget https://nginx.org/download/nginx-1.24.0.tar.gz
```

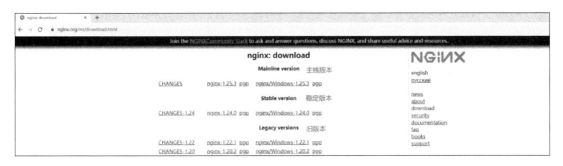

图 5　Nginx 官网下载

3. 解压 Nginx。解压下载的 Nginx 源代码包。

```
tar -zxvf nginx-1.24.0.tar.gz
```

4. 编译和安装。进入解压后的 Nginx 目录并进行编译和安装。

```
# 切换到 Nginx 解压目录
cd nginx-1.24.0
# 编译前的配置和依赖检查
./configure
# 编译安装
make && make install
```

Nginx 安装完成后,默认自动创建/usr/local/nginx 目录,并创建必要的文件和目录,包

括配置文件、日志文件、HTML 文件等。

5. 防火墙设置。如果系统启用了防火墙,需要关闭防火墙。

查看防火墙状态
systemctl status firewalld

关闭防火墙
systemctl stop firewalld

开机禁用防火墙
systemctl disable firewalld

6. 启动 Nginx,进入 Nginx 的安装目录。

cd /usr/local/nginx/sbin

进入目录后,执行下方指令,运行 Nginx。

./nginx

Nginx 启动成功后,默认运行在 80 端口,可以用浏览器打开自己服务器 IP 地址,查看页面效果,如图 6 所示。

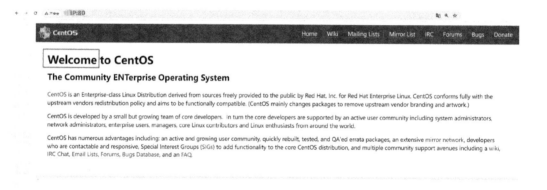

图 6　Nginx 启动成功

3.2.2　配置线上接口

我们之前配置的接口为本地接口,现在部署线上服务器,需要将之前的接口改为线上地址,需要在 vite.config.ts 文件中修改 target 地址为自己后端部署地址。

```
server: {
    host: '0.0.0.0',
    port: Number(env.VITE_PORT),
```

```
open: env. VITE_OPEN = = = 'true',
// 设置 server. hmr. overlay 为 false 可以屏蔽开发服务器中的错误提示
// hmr: {
//    overlay: false
// },
cors: true,
// 跨域代理配置
proxy: {
  "/dev": {
    target: "后端线上地址",
    changeOrigin: true,
    rewrite: path => path. replace(/^\/dev/, "")
  }
}
}
}
```

3.2.3 项目打包

进入后台管理系统项目（图7），进行打包操作，执行命令，查看项目文件中新生成的 dist 文件夹（图8）。

图7 系统打包成功效果图

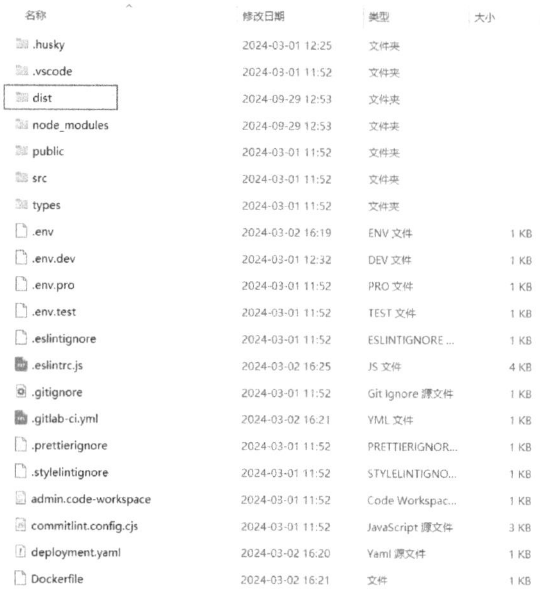

图 8 查看项目文件中新生成的 dist 文件夹

3.2.4 复制打包文件到 Nginx

安装好 Nginx，找到安装根目录。如图 9 所示，将 dist 目录里的文件复制到服务器 Nginx 文件夹的 html 目录下。这是默认的 nginx 的根目录，这样做不用修改配置文件就可以直接运行，查看页面效果即可。

图 9 复制打包文件到 Nginx 文件夹 html 目录下

复制好对应文件后,网页打开自己的服务器 80 端口即可看到部署之后的页面效果(图 10)。

图 10　部署成功效果图

4　任务总结

在本次前后端分离项目的部署过程中,我们重点关注了环境搭建和配置工作,确保项目的运行稳定和高效。

后端方面,我们在 Linux 系统上安装了 JDK 17、MySQL 8.0 和 Redis 等关键软件,为后端 API 提供了稳定的基础运行环境。这些软件的安装和配置确保了后端服务能够高效地处理请求和数据存储,为项目的顺利运行奠定了坚实基础。

前端部分,我们基于 Vue 3 框架进行开发,并通过 Vite 进行打包优化。最终,前端应用被部署到 Nginx 服务器上。Nginx 作为一款高效的文件处理和反向代理服务器,能够快速地加载前端资源,并通过反向代理实现与后端服务的顺畅通信。这一部署策略不仅提升了用户体验,也确保了前后端之间的紧密集成和高效协作。

结束任务

客户端系统

波哥召集了项目开发团队:"这段时间大家辛苦了,我们资源分享应用后台管理系统已经顺利完成了。接下来,我们的重点将转向客户端系统的开发,这主要包括客户端小程序的开发和客户端 API 的对接。"

小南:"波哥,我明白了。但是,客户端小程序和 API 的开发这部分内容,好像没有在前面的培训和开发中涉及过。"

波哥回答道:"你说得对!这部分内容相对独立且深入,所以我们将其列入下一个开发计划。"

小工:"那我们接下来该怎么做呢?是直接开始开发吗,还是需要先学习下开发计划中的内容?"

波哥:"我们当然需要先熟悉一下开发计划中的内容,至少要了解小程序的基本架构和 API 的开发规范。不过,我建议大家可以先从一些基础的开发工作开始,比如搭建小程序的开发环境,梳理 API 的接口文档等。等大家掌握了客户端小程序和 API 的相关技术后,再来启动下一轮的客户端系统开发计划。"

小南:"明白了,我可以先开始搭建小程序的开发环境,同时学习小程序的开发文档。"

小工:"那我这边就先开始梳理 API 的接口文档,并准备好开发环境。"

波哥:"很好,你们两个分工明确,这样效率会更高。如果遇到什么问题或者需要进一步的帮助,记得随时找我或者团队其他成员。记住,我们的目标是顺利完成客户端系统的开发,为项目画上圆满的句号。加油!"

小南、小工齐声道:"好的,波哥!我们会努力的!"

任务点

- 了解客户端系统整体功能;
- 知道最新主流的客户端系统技术选型;
- 了解客户端系统开发规划。

任务计划

- 任务内容:了解客户端系统的功能,技术选型和开发规划等;
- 任务耗时:预计完成时间为 30 min ~ 1 h;
- 任务难点:分析客户端系统的技术选型。

1 客户端整体功能

客户端功能如表 1 所示。

表 1 客户端模块对应的功能

模块	功能
账户模块	注册登录；绑定/解绑手机号；用户信息显示/修改；用户积分查询；收藏、签到
首页模块	指定通知滚动；首页通知轮播；资源分类查询展示；热门标签
投稿模块	用户投稿
搜索模块	根据关键词、热门标签、搜索记录进行搜索

2 客户端技术选型

客户端后端 API 主要技术选型如表 2 所示。

表 2 客户端后端 API 主要技术选型

名称	作用
Spring Boot	后端整体框架
MySQL	数据库
MyBatis Plus	数据持久层
Redis	缓存
阿里云 OSS SDK	对象存储
容联云 SDK	第三方短信
mapstruct	实体类映射
knife4j	在线接口文档
hutool	常用工具包

客户端小程序主要技术选型如表 3 所示。

表3　客户端小程序主要技术选型

名称	作用
uni-app	前端整体框架
uni-ui	UI 库
pinia	状态管理
ES Lint	代码规则
prettier	代码格式化

3　项目开发规划

客户端的整体开发规划如表4所示。

表4　客户端整体开发规划

阶段任务	细分任务
客户端 API 开发	任务一:项目介绍
	任务二:设计数据库
	任务三:设计接口文档
	任务四:搭建客户端 API 项目
	任务五:开发基础服务接口
	任务六:开发认证接口
	任务七:开发用户接口
	任务八:开发公告接口
	任务九:开发标签分类接口
	任务十:开发资源接口
	任务十一:API 项目打包部署
客户端小程序开发	任务一:搭建客户端小程序项目
	任务二:开发账户模块
	任务三:开发首页模块
	任务四:开发资源模块
	任务五:开发投稿模块
	任务六:开发个人中心模块
	任务七:开发搜索模块
	任务八:小程序项目打包

4　任务总结

本次任务主要是了解客户端系统的整体功能,包括客户端的基础功能、客户端技术选型和开发计划,理解项目的技术选型,了解项目的开发规划。

客户端系统作为项目的重要组成部分,承担着与用户直接交互的任务。其中,客户端小程序的开发和客户端 API 的对接是客户端系统开发的两大核心内容。然而,由于篇幅和内容的限制,本教材并未对这两部分内容进行详细的讲解。但请读者放心,关于客户端系统的教程,我们将在另一本教材中进行全面而深入的介绍。

在另一本教材中,我们将重点介绍客户端系统的开发流程、技术选型、架构设计、功能实现等方面的内容。针对小程序的开发,我们将从环境搭建、开发规范、页面设计、功能实现等多个角度进行详细介绍,帮助读者快速掌握小程序的开发技能。同时,针对客户端 API 的对接,我们将从接口设计、数据传输、安全性保障等方面进行深入剖析,确保读者能够顺利地实现前后端的数据交互。

为了顺利完成客户端系统的开发工作,我们建议读者在掌握本教材内容的基础上,尽快熟悉另一本教材的内容。在学习的过程中,可以结合项目实际需求,有针对性地进行学习和实践。同时,我们鼓励读者积极参与讨论和交流,共同解决在开发过程中遇到的问题。

总之,通过本教材的学习和另一本教材的深入实践,读者将能够全面掌握软件系统的开发技能,为项目的成功实施奠定坚实的基础。我们期待在未来的学习和工作中,与读者一起成长和进步。